Studies in Fuzziness and Soft Computing

Volume 351

Series editor

Janusz Kacprzyk, Polish Academy of Sciences, Warsaw, Poland
e-mail: kacprzyk@ibspan.waw.pl

About this Series

The series "Studies in Fuzziness and Soft Computing" contains publications on various topics in the area of soft computing, which include fuzzy sets, rough sets, neural networks, evolutionary computation, probabilistic and evidential reasoning, multi-valued logic, and related fields. The publications within "Studies in Fuzziness and Soft Computing" are primarily monographs and edited volumes. They cover significant recent developments in the field, both of a foundational and applicable character. An important feature of the series is its short publication time and world-wide distribution. This permits a rapid and broad dissemination of research results.

More information about this series at http://www.springer.com/series/2941

Krassimir T. Atanassov

Intuitionistic Fuzzy Logics

 Springer

Krassimir T. Atanassov
Department of Bioinformatics and
   Mathematical Modelling, Institute of
   Biophysics and Biomedical Engineering
Bulgarian Academy of Sciences
Sofia
Bulgaria

ISSN 1434-9922   ISSN 1860-0808   (electronic)
Studies in Fuzziness and Soft Computing
ISBN 978-3-319-84055-0   ISBN 978-3-319-48953-7   (eBook)
DOI 10.1007/978-3-319-48953-7

Printed on acid-free paper

This Springer imprint is published by Springer Nature
The registered company is Springer International Publishing AG
The registered company address is: Gewerbestrasse 11, 6330 Cham, Switzerland

In memory of George Gargov

Preface

The work over the present book started in 1993 in a discussion with George Gargov. Ten years earlier, he had proposed the name of the extension of fuzzy sets, introduced by me in [1]. In the next years, first I and later together with G. Gargov gave some of the basic definitions of intuitionistic fuzzy propositional calculus, intuitionistic fuzzy predicate logic and intuitionistic fuzzy modal logic. Unfortunately, in 1996, Gargov died. During next 20 years, I was gathering enthusiasm to write this book. Meantime, I published about 30 books in the areas of the generalized nets (extensions of the Petri nets), intuitionistic fuzzy sets and number theory, but the work on the present book was progressing very slowly. The readers will see that there are a lot of open problems in the book, which I cannot solve at the moment, but I hope that the readers will be more successful than me in this enterprise.

I will repeat a paragraph of my book [2], that I think will be valid for the present book, too: "throughout the years, many colleagues of mine have shown me or published communications about inaccuracies, found in [3]. I myself have also found many mistakes, most of which were due to the "copy-paste" effect. I do not cherish any illusions that the present book is perfect. However, I do believe that the present book holds some interesting ideas that open opportunities for future research. I would encourage researchers to pay more attention in this direction."

During last years, in the area of fuzzy logic a lot of books have been published, which had an impact on my research. An incomprehensive list includes [4–17]. To the best of my knowledge, this book is the first one, exclusively devoted to intuitionistic fuzzy logics.

Now, I have recorded more than 5000 papers on intuitionistic fuzzy sets, written by researchers from more than 50 countries. Most of these papers discuss important issues related to the theory or applications of the intuitionistic fuzzy sets. For this reason, I realized that it would be virtually impossible to make a survey of all the others' results and contented myself with the presentation of my own works only. I am convinced that in recent future many of the colleagues in the field will likewise collect and publish their own ideas as papers and monographs, and I would gladly subserve anyone in that.

I am very thankful to my colleagues and students who motivated me to prepare the present book. The list of their names is long, but I must note (in the alphabetical order of their first names) at least those of my coauthors whose research I have used and included in this book: Beloslav Riečan, Boyan Kolev, Dimiter Dimitrov, Eulalia Szmidt, Evgeniy Marinov, Ivan Georgiev, Janusz Kacprzyk, Lilija Atanassova, Nora Angelova, Peter Vassilev, Radoslav Tsvetkov, Trifon Trifonov, Vassia Atanassova.

Ivan Georgiev, Peter Vassilev and my daughter Vassia Atanassova read and corrected the text of the manuscript and deserve my deepest thanks and appreciation.

Special thanks are due to one of the most active contributors and promoters of intuitionistic fuzzy sets worldwide, an active researcher, coauthor and friend–Janusz Kacprzyk, the Editor-in-chief of this book series, who urged me to finalize my 30-year-long work on intuitionistic fuzzy logics.

I am grateful for the support provided by the project Ref. No. DFNI-I-02-5 funded by the Bulgarian Science Fund.

Last but not least, I would like to thank the three most important ladies in my life: my mother, my wife, and my daughter for encouraging and stimulating all my scientific research.

Sofia, Bulgaria   Krassimir T. Atanassov
July 2016

References

1. Atanassov, K. Intuitionistic fuzzy sets, VII ITKR's Session, Sofia, June 1983 (Deposed in Central Sci. - Techn. Library of Bulg. Acad. of Sci., 1697/84) (in Bulg.), Reprinted: Int J Bioautomation, 2016;2(S1):S1–S6.
2. Atanassov, K. On Intuitionistic Fuzzy Sets Theory, Springer, Berlin, 2012.
3. Atanassov, K. Intuitionistic Fuzzy Sets, Springer, Heidelberg, 1999.
4. Aliev, R. A., Fundamentals of the Fuzzy Logic-Based Generalized Theory of Decisions. Springer, Berlin, 2013.
5. Baczynski, M., G. Beliakov, H. Bustince and A. Pradera (Eds.), Advances in Fuzzy Implication Functions, Springer, Berlin, 2013.
6. Baczynski, M., B. Jayaram, Fuzzy Implications, Springer, Berlin, 2008.
7. Bede, B., Mathematics of Fuzzy Sets and Fuzzy Logic, Springer, Berlin, 2013.
8. Belohlavek, R., V. Vychodil, Fuzzy Equational Logic, Springer, Berlin, 2005.
9. Castillo, O., Type-2 Fuzzy Logic in Intelligent Control Applications, Springer, Berlin, 2012.
10. Castillo, O., P. Melin, Type-2 Fuzzy Logic: Theory and Applications, Springer, Berlin, 2008.
11. Klir, G., B. Yuan, Fuzzy Sets and Fuzzy Logic. Prentice Hall, New Jersey, 1995.
12. Orlowska, E. (Ed.), Logic at Work, Springer, Heidelberg, 1999.
13. Rasiowa, H., An Algebraic Approach to Non-Classical Logics, Polish Scientific Publishers, Warszawa, 1974.
14. Stankovic, R., J. Astola, From Boolean Logic to Switching Circuits and Automata, Springer, Berlin, 2011.
15. Szmidt, E., Distances and Similarities in Intuitionistic Fuzzy Sets, Springer, Berlin, 2014.
16. Wang, P., D. Ruan, E. Kerre (Eds.), Fuzzy Logic, Springer, Berlin, 2007.
17. Xu, Y., D. Ruan, K. Qin, J. Liu, Lattice-Valued Logic, Springer, Berlin, 2003.

Contents

1 Elements of Intuitionistic Fuzzy Propositional Calculus   1
   1.1   Intuitionistic Fuzzification of the Validity of Propositions   1
   1.2   Intuitionistic Fuzzy Implications   9
   1.3   Discussion on Intuitionistic Fuzzy Implications   10
   1.4   Intuitionistic Fuzzy Negations   18
   1.5   Properties of Intuitionistic Fuzzy Implications and Negations   19
   1.6   De Morgan Laws and Law for Excluded Middle   52
   1.7   New Intuitionistic Fuzzy Conjunctions and Disjunctions   56
   References   60

2 Intuitionistic Fuzzy Predicate Logic   65
   2.1   Short Remarks on Intuitionistic Fuzzy Predicate Logic   65
   2.2   Extended Intuitionistic Fuzzy Quantifiers   71
   2.3   Ideas for New Types of Quantifiers   75
   References   76

3 Intuitionistic Fuzzy Modal Logics   79
   3.1   Intuitionistic Fuzzy Classical Modal Operators   79
   3.2   Extensions of the Intuitionistic Fuzzy Modal Operators   93
   3.3   Second Type of Intuitionistic Fuzzy Modal Operators   106
   3.4   Intuitionistic Fuzzy Level Operators   120
   3.5   Pseudo-fixed Points of the Intuitionistic Fuzzy Operators
         and Quantifiers   122
   References   123

4 Temporal and Multidimensional Intuitionistic Fuzzy Logics   125
   4.1   Temporal Intuitionistic Fuzzy Logic   125
   4.2   Multidimensional Intuitionistic Fuzzy Logics   129
   References   133

5 Conclusion   135

Index   137

Chapter 1
Elements of Intuitionistic Fuzzy Propositional Calculus

1.1 Intuitionistic Fuzzification of the Validity of Propositions

In classical logic (e.g., [1–4]), to each proposition (sentence) we juxtapose its truth value: truth – denoted by 1, or falsity – denoted by 0. In the case of fuzzy logic [5], this truth value is a real number in the interval [0, 1] and it is called "truth degree" or "degree of validity". In the intuitionistic fuzzy case (see [6–9]) we add one more value – "falsity degree" or "degree of non-validity"– which is again in interval [0, 1]. Thus, to the proposition p, two real numbers, $\mu(p)$ and $\nu(p)$, are assigned with the following constraint:

$$\mu(p), \nu(p) \in [0, 1] \text{ and } \mu(p) + \nu(p) \le 1. \tag{1.1.1}$$

Let

$$\pi(p) = 1 - \mu(p) - \nu(p).$$

This function determines the degree of uncertainty (indeterminacy).

In [10], the pair $\langle \mu(p), \nu(p) \rangle$ that satisfies condition (1.1.1) is called "Intuitionistic Fuzzy Pair" (IFP).

Let an evaluation function V be defined over a set of propositions $\mathcal{S}$, in such a way that for $p \in \mathcal{S}$:

$$V(p) = \langle \mu(p), \nu(p) \rangle.$$

Hence the function $V : \mathcal{S} \to [0, 1] \times [0, 1]$ gives the truth and falsity degrees of all elements of $\mathcal{S}$.

We assume that the evaluation function V assigns to the logical truth T

$$V(T) = \langle 1, 0 \rangle,$$

© Springer International Publishing AG 2017
K.T. Atanassov, *Intuitionistic Fuzzy Logics*, Studies in Fuzziness and Soft Computing 351, DOI 10.1007/978-3-319-48953-7_1

and to the logical falsity F

$$V(F) = \langle 0, 1 \rangle.$$

When

$$\nu(p) = 1 - \mu(p),$$

i.e.,

$$V(p) = \langle \mu(p), 1 - \mu(p) \rangle,$$

it coincides with the fuzzy case.

As it was discussed in the author's book [11], one of his major mistakes was that in the middle of 1980s, when he found the following two negations that can be defined over elements of S, he did not study in details the properties of the second negation, because it was essentially more complex. For $p \in S$ these negations are:

$$V(\neg_1 p) = \langle \nu(p), \mu(p) \rangle, \tag{1.1.2}$$

$$V(\neg_2 p) = \langle \overline{sg}(\mu(p)), sg(\mu(p)) \rangle, \tag{1.1.3}$$

where here and below

$$sg(x) = \begin{cases} 1 & \text{if } x > 0 \\ 0 & \text{if } x \leq 0 \end{cases},$$

and

$$\overline{sg}(x) = \begin{cases} 0 & \text{if } x > 0 \\ 1 & \text{if } x \leq 0 \end{cases}.$$

Obviously, the first definition coincides with the formula in the ordinary fuzzy logic (see, e.g., [12]).

Immediately, we see that for each proposition $p \in S$:

$$V(\neg_1 \neg_1 p) = V(p),$$

which some colleagues interpreted as a contradiction with the idea of L. Brouwer's intuitionism (see, e.g., [13–15]). The author's opinion is that the words "intuitionistic fuzzy" correspond to the form of the elements of set S, but not to the forms of the operations defined over the intuitionistic fuzzy propositions.

The second negation does not satisfy the equality $V(\neg_2 \neg_2 p) = V(p)$, and as we see below, it exhibits truly intuitionistic behaviour. This is a confirmation of the author's assertion that intuitionistic fuzzy objects have intuitionistic properties. In the Sect. 1.4, we discuss not only the second, but more than 50 different negations. Here, we define only the operations "disjunction", "conjunction" and "implication", originally introduced in [6], that have classical logic analogues, as follows:

$$V(p \vee q) = \langle \max(\mu(p), \mu(q)), \min(\nu(p), \nu(q)) \rangle, \qquad (1.1.4)$$
$$V(p \wedge q) = \langle \min(\mu(p), \mu(q)), \max(\nu(p), \nu(q)) \rangle, \qquad (1.1.5)$$
$$V(p \rightarrow q) = \langle \max(\nu(p), \mu(q)), \min(\nu(p), \mu(q)) \rangle. \qquad (1.1.6)$$

In some places below, we call them "standard" disjunction, conjunction and implication.

Similarly to [7, 16], several geometrical interpretations of the results of the function V will be discussed below.

In intuitionistic fuzzy propositional calculus, the formulas are defined in the manner of standard propositional calculus.

In [7], it is noted that the ordinary fuzzy sets have only one geometrical interpretation, while in [7, 16] several interpretations of IFSs are given. Here we show the most relevant interpretations for the logical case.

The first one (which is analogous to the standard fuzzy set interpretation) is shown on Fig. 1.1.

Its analogue is given in Fig. 1.2.

Fig. 1.1 First geometrical interpretation

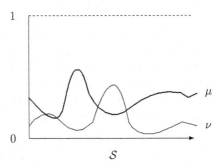

Fig. 1.2 Modified form of the first geometrical interpretation

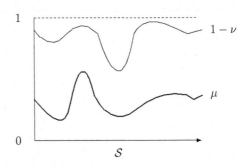

Therefore, we can map to every formula A, a unit segment in the form:

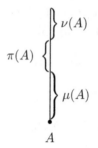

Let a universe $\mathcal{S}$ be given and let us consider the figure F in the Euclidean plane with a Cartesian coordinate system (see Fig. 1.3). Then, we can construct an evaluation function V from $\mathcal{S}$ to F such that if $p \in \mathcal{S}$, then

$$x = V(p) \in F,$$

the point x has coordinates $\langle a, b \rangle$ for which: $0 \le a + b \le 1$ and these coordinates are such that $a = \mu(p), b = \nu(p)$.

We will note that there can exist two different elements $p, q \in \mathcal{S}$ for which $\mu(p) = \mu(q)$ and $\nu(p) = \nu(q)$, i.e., for which $V(p) = V(q)$.

About the form and the methods of determining the functions μ and ν, we must repeat the same, as in [7]: everywhere below we will assume that these functions are either pre-determined or obtained as a result of the application of some operations or operators over pre-determined functions. In fuzzy set theory, there are three basic ways to construct membership functions:

(i)   on the basis of expert knowledge;
(ii) explicitly—on the basis of observations collected in advance and processed appropriately (e.g., by statistical methods);
(iii) analytically—by suitably chosen functions (e.g. probabilistic distribution).

Fig. 1.3 Second geometrical interpretation

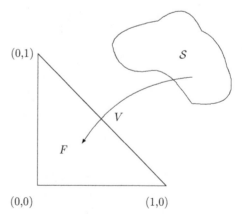

The two latter cases are treated in much the same way as ordinary fuzzy sets; however these methods are now used for the estimation of both the degree of membership and the degree of non-membership of a given element of a fixed universe to a subset of the same universe. It is clear that a correct method must respect the inequalities

$$0 \leq \mu(A) + \nu(A) \leq 1$$

for every formula A.

Following [7], we must also add, that the case when the functions values are calculated on the basis of expert knowledge is more complicated. In this case problems related to the correctness of the expert estimations arise. No such problems arise when dealing with ordinary fuzzy sets. These problems are discussed in Sect. 4.3 of [7], where five approaches of processing expert knowledge are proposed, regarding the construction of the degrees of membership and non-membership. These approaches are introduced in increasing order of complexity, and they reflect the assurance of the experts who estimate the corresponding events (objects, processes), their personal and collective opinion and their expert ratings. Similar approaches can be used for processing collected knowledge (observations), when incorrect data are suspected. Some of the methods from Sect. 4.3 of [7] and the methods introduced by P. Dworniczak in [17, 18], can help us locate the incorrect pieces of information.

Four other geometrical interpretations are shown in Figs. 1.4, 1.5 and 1.6.

The triangle from Fig. 1.4 has sides length of $\frac{2\sqrt{3}}{3}$ and therefore, the length of the altitude is equal to $1 \; (= \mu(p) + \nu(p) + \pi(p))$.

The angles to the basis of the triangle from Fig. 1.5 have the values, respectively, $\alpha = \pi.\mu(p)$ and $\beta = \pi.\nu(p)$, where $\pi = 3.14159\ldots$, and the length of the triangle basis is equal to 1, as well as the catheti of the rectangular triangle from Fig. 1.6.

In [19], a geometrical interpretation based on radar chart is proposed by V. Atanassova. In Fig. 1.7, the innermost zone corresponds to the membership degree, the outermost zone to the non-membership degree and the region between both zones to the degree of uncertainty. This interpretation can be especially useful for data in time series, multivaried data sets and other data with cyclic trait.

Fig. 1.4 Third geometrical interpretation

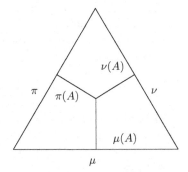

Fig. 1.5 Fourth geometrical interpretation

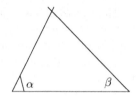

Fig. 1.6 Fifth geometrical interpretation

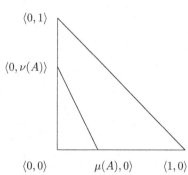

Fig. 1.7 Sixth geometrical interpretation

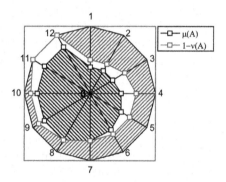

The way the two functions, μ and ν, are constructed will not be important for our considerations below.

The geometrical interpretations of the first two operations conjunction and disjunction are the following.

Let p and q be two propositions in S. Let the evaluation function V assign to $p \wedge q \in S$ a point $V(p \wedge q) \in F$ with coordinates

$$\langle \min(\mu(p), \mu(q)), \max(\nu(p), \nu(q)) \rangle.$$

There exist three geometrical cases (see Fig. 1.8a–c) - one general case (Fig. 1.8a) and two particular cases (Fig. 1.8b, c).

Now, the evaluation function V assigns to $p \vee q \in S$ a point $V(p \vee q) \in F$ with coordinates

$$\langle \max(\mu(p), \mu(q)), \min(\nu(p), \nu(q)) \rangle.$$

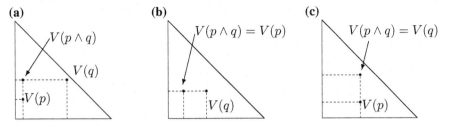

Fig. 1.8 Second geometrical interpretation of operation $\wedge$

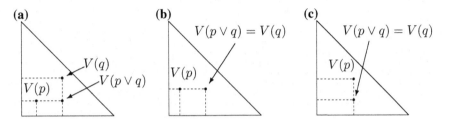

Fig. 1.9 Second geometrical interpretation of operation $\vee$

There exist also three geometrical cases as above, namely, one general case (Fig. 1.9a) and two particular cases (Fig. 1.9b, c).

In 1988 two different implications were defined (see [6, 16]). For the case of the first implication (the classical one, see (1.1.6)), function V assigns to $p \rightarrow q \in S$ a point $V(p \rightarrow q) \in F$ with coordinates $\langle \max(b, c), \min(a, d) \rangle$ (Fig. 1.10a–h).

The second implication (non-classical) has the form

$$V(p \rightarrow q) = \langle 1 - (1 - c.\mathrm{sg}(a - c)), d.\mathrm{sg}(a - c)\mathrm{sg}(d - b) \rangle.$$

For the case of the second implication, evaluation function V assigns to $p \rightarrow q \in S$ a point $V(p \rightarrow q) \in F$ with coordinates $\langle 1 - (1 - c.\mathrm{sg}(a-c)), d.\mathrm{sg}(a-c)\mathrm{sg}(d-b) \rangle$ (Fig. 1.11a–h).

In the beginning of this century, when the author started re-defining the multiple fuzzy implications in intuitionistic fuzzy form, he saw that the above implication coincides with implication $\rightarrow_3$ that we will discuss in the next section.

All this is valid for formulas, instead of propositions, too.

For the needs of the discussion below, we define the notions of Intuitionistic Fuzzy Tautology (IFT, see, e.g. [6, 11]) and tautology.

Formula A is an IFT if and only if (iff) for every evaluation function V, if $V(A) = \langle a, b \rangle$, then,

$$a \geq b, \tag{1.1.7}$$

while it is a (classical) tautology if and only if for every evaluation function V, if $V(A) = \langle a, b \rangle$, then,

$$a = 1, \ b = 0. \tag{1.1.8}$$

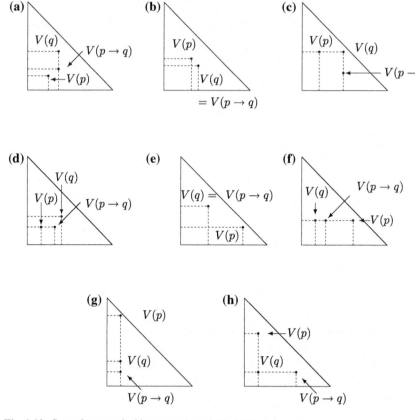

Fig. 1.10 Second geometrical interpretation of the classical operation →

For each evaluation function V and for each formula A such that $V(x) = \langle a, b \rangle$, let us say that A is "intuitionistic fuzzy sure" (IF-sure), iff $a \geq 0.5 \geq b$.

Below, when it is clear, we will omit notation "$V(A)$", using directly "A" of the intuitionistic fuzzy evaluation of A. Also, for brevity, in a lot of places, instead of the IFP $\langle \mu(A), \nu(A) \rangle$ we will use the IFP $\langle a, b \rangle$, where $a, b \in [0, 1]$ and $a + b \leq 1$.

It is also suitable, if $\langle a, b \rangle$ and $\langle c, d \rangle$ are IFPs, to have

$$\langle a, b \rangle \leq \langle c, d \rangle \ \text{ iff } \ a \leq c \ \text{ and } \ b \geq d \tag{1.1.9}$$

and

$$\langle a, b \rangle \geq \langle c, d \rangle \ \text{ iff } \ a \geq c \ \text{ and } \ b \leq d. \tag{1.1.10}$$

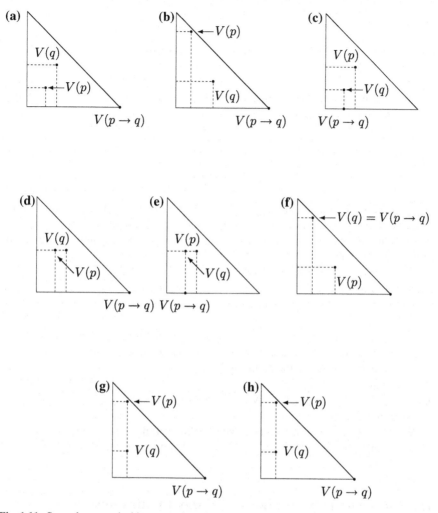

Fig. 1.11 Second geometrical interpretation of the first non-classical operation $\rightarrow$

1.2 Intuitionistic Fuzzy Implications

In the series of papers [20–55], 185 different implications were defined and some of their basic properties were studied. In some of these publications, some misprints in the formulas were found during the last years. Here, we give the full list of the corrected intuitionistic fuzzy implications.

The first 10 implications, described in [23] are intuitionistic fuzzy analogues of existing fuzzy implications in literature (see, e.g., [56]). The next five implications were introduced in author's publications [24, 31, 36, 37]. Two of them are given in papers of B. Kolev [31] ($\rightarrow_{13}$) and of T. Trifonov ($\rightarrow_{14}$) [36, 37] and the author.

From the so constructed 15 implications, using the formula

$$\neg A = A \to F, \tag{1.2.1}$$

five negations are formulated – the standard (classical) negation $\neg_1$ and four others $\neg_2, \ldots, \neg_5$, where F is defined above and $\to$ is each one of these 15 implications [57]. By formulas

$$A \to B = \neg A \lor B$$

and

$$A \to B = \neg A \lor \neg\neg B,$$

these 5 negations generate eight new implications [25]. Each of these implications generates four new implications, by formulas using intuitionistic fuzzy (standard, classical) modal operators, that we discuss in Chap. 3. The analogues of these implications are given in [11].

In [42–45], L. Atanassova introduced 11 new implications ($\to_{139}, \ldots, \to_{149}$). P. Dworniczak generalized them in [52–54] (implications $\to_{150}, \ldots, \to_{152}$) and L. Atanassova modified Dworniczak's implications in [49–51] (implications $\to_{154}, \ldots, \to_{165}$).

The next five implications were introduced by the author in [58] as modifications of the first L. Zadeh's type intuitionistic fuzzy implication $\to_1$, implications $\to_{171}, \to_{175}$ were introduced by the author in [29].

Implications $\to_{176}, \ldots, \to_{180}$ were proposed by E. Szmidt, J. Kacprzyk and the author in [33–35], while the last five implications $\to_{181}, \ldots, \to_{185}$ were introduced by the author in [29].

All currently existing implications were published in [29] and they are given in Table 1.1. In it, we keep the numeration from [11] up to number 138.

1.3 Discussion on Intuitionistic Fuzzy Implications

Let us have

$$\lambda \geq 1, \ \gamma \geq 1, \ \alpha \geq 1, \beta \in [1, \alpha],$$

$$\varepsilon, \eta \in [0, 1] \text{ and } \varepsilon \leq \eta < 1.$$

By direct check, we see that the following equalities hold.

Table 1.1 List of the intuitionistic fuzzy implications

$\rightarrow_1$	$\langle \max(b, \min(a, c)), \min(a, d) \rangle$
$\rightarrow_2$	$\langle \overline{sg}(a - c), d\,sg(a - c) \rangle$
$\rightarrow_3$	$\langle 1 - (1 - c)sg(a - c)), d\,sg(a - c) \rangle$
$\rightarrow_4$	$\langle \max(b, c), \min(a, d) \rangle$
$\rightarrow_5$	$\langle \min(1, b + c), \max(0, a + d - 1) \rangle$
$\rightarrow_6$	$\langle b + ac, ad \rangle$
$\rightarrow_7$	$\langle \min(\max(b, c), \max(a, b), \max(c, d)),$ $\max(\min(a, d), \min(a, b), \min(c, d))) \rangle$
$\rightarrow_8$	$\langle 1 - (1 - \min(b, c))sg(a - c), \max(a, d)sg(a - c), sg(d - b) \rangle$
$\rightarrow_9$	$\langle b + a^2 c, ab + a^2 d \rangle$
$\rightarrow_{10}$	$\langle c\overline{sg}(1 - a) + sg(1 - a)(\overline{sg}(1 - c) + bsg(1 - c)),$ $d\overline{sg}(1 - a) + asg(1 - a)sg(1 - c)) \rangle$
$\rightarrow_{11}$	$\langle 1 - (1 - c)sg(a - c), d\,sg(a - c)sg(d - b) \rangle$
$\rightarrow_{12}$	$\langle \max(b, c), 1 - \max(b, c) \rangle$
$\rightarrow_{13}$	$\langle b + c - bc, ad \rangle$
$\rightarrow_{14}$	$\langle 1 - (1 - c)sg(a - c) - d\overline{sg}(a - c)sg(d - b), d\,sg(d - b) \rangle$
$\rightarrow_{15}$	$\langle 1 - (1 - \min(b, c))sg(sg(a - c) + sg(d - b))$ $- \min(b, c)sg(a - c)sg(d - b),\ 1 - (1 - \max(a, d))sg(\overline{sg}(a - c)$ $+ \overline{sg}(d - b)) - \max(a, d)\overline{sg}(a - c)\overline{sg}(d - b) \rangle$
$\rightarrow_{16}$	$\langle \max(\overline{sg}(a), c), \min(sg(a), d) \rangle$
$\rightarrow_{17}$	$\langle \max(b, c), \min(ab + a^2, d) \rangle$
$\rightarrow_{18}$	$\langle \max(b, c), \min(1 - b, d) \rangle$
$\rightarrow_{19}$	$\langle \max(1 - sg(sg(a) + sg(1 - b)), c), \min(sg(1 - b), d) \rangle$
$\rightarrow_{20}$	$\langle \max(\overline{sg}(a), sg(c)), \min(sg(a), \overline{sg}(c)) \rangle$
$\rightarrow_{21}$	$\langle \max(b, c(c + d)), \min(a(a + b), d(c^2 + d + cd)) \rangle$
$\rightarrow_{22}$	$\langle \max(b, 1 - d), 1 - \max(b, 1 - d) \rangle$
$\rightarrow_{23}$	$\langle 1 - \min(sg(1 - b), \overline{sg}(1 - d)), \min(sg(1 - b), \overline{sg}(1 - d)) \rangle$
$\rightarrow_{24}$	$\langle \overline{sg}(a - c)\overline{sg}(d - b), sg(a - c)sg(d - b) \rangle$
$\rightarrow_{25}$	$\langle \max(b, \overline{sg}(a)\overline{sg}(1 - b)), c\overline{sg}(d)\overline{sg}(1 - c)), \min(a, d) \rangle$
$\rightarrow_{26}$	$\langle \max(\overline{sg}(1 - b), c), \min(sg(a), d) \rangle$
$\rightarrow_{27}$	$\langle \max(\overline{sg}(1 - b), sg(c)), \min(sg(a), \overline{sg}(1 - d)) \rangle$
$\rightarrow_{28}$	$\langle \max(\overline{sg}(1 - b), c), \min(a, d) \rangle$
$\rightarrow_{29}$	$\langle \max(\overline{sg}(1 - b), \overline{sg}(1 - c)), \min(a, \overline{sg}(1 - d)) \rangle$
$\rightarrow_{30}$	$\langle \max(1 - a, \min(a, 1 - d)), \min(a, d) \rangle$
$\rightarrow_{31}$	$\langle \overline{sg}(a + d - 1), d\,sg(a + d - 1) \rangle$
$\rightarrow_{32}$	$\langle 1 - d\,sg(a + d - 1), d\,sg(a + d - 1) \rangle$
$\rightarrow_{33}$	$\langle 1 - \min(a, d), \min(a, d) \rangle$
$\rightarrow_{34}$	$\langle \min(1, 2 - a - d), \max(0, a + d - 1) \rangle$

(continued)

Table 1.1 (continued)

$\rightarrow_{35}$	$\langle 1 - ad, ad \rangle$
$\rightarrow_{36}$	$\langle \min(1 - \min(a, d), \max(a, 1 - a), \max(1 - d, d)), \max(\min(a, d),$ $\min(a, 1 - a), \min(1 - d, d)) \rangle$
$\rightarrow_{37}$	$\langle 1 - \max(a, d)\text{sg}(a + d - 1), \max(a, d)\text{sg}(a + d - 1) \rangle$
$\rightarrow_{38}$	$\langle 1 - a + a^2(1 - d), a(1 - a) + a^2 d \rangle$
$\rightarrow_{39}$	$\langle (1 - d)\overline{\text{sg}}(1 - a) + \text{sg}(1 - a)(\overline{\text{sg}}(d) + (1 - a)\text{sg}(d)),$ $d\overline{\text{sg}}(1 - a) + a\text{sg}(1 - a)\text{sg}(d) \rangle$
$\rightarrow_{40}$	$\langle 1 - \text{sg}(a + d - 1), 1 - \overline{\text{sg}}(a + d - 1) \rangle$
$\rightarrow_{41}$	$\langle \max(\overline{\text{sg}}(a), 1 - d), \min(\text{sg}(a), d) \rangle$
$\rightarrow_{42}$	$\langle \max(\overline{\text{sg}}(a), \text{sg}(1 - d)), \min(\text{sg}(a), \overline{\text{sg}}(1 - d)) \rangle$
$\rightarrow_{43}$	$\langle \max(\overline{\text{sg}}(a), 1 - d), \min(\text{sg}(a), d) \rangle$
$\rightarrow_{44}$	$\langle \max(\overline{\text{sg}}(a), 1 - d), \min(a, d) \rangle$
$\rightarrow_{45}$	$\langle \max(\overline{\text{sg}}(a), \overline{\text{sg}}(d)), \min(a, \overline{\text{sg}}(1 - d)) \rangle$
$\rightarrow_{46}$	$\langle \max(b, \min(1 - b, c)), 1 - \max(b, c) \rangle$
$\rightarrow_{47}$	$\langle \overline{\text{sg}}(1 - b - c), (1 - c)\text{sg}(1 - b - c) \rangle$
$\rightarrow_{48}$	$\langle 1 - (1 - c)\text{sg}(1 - b - c), (1 - c)\text{sg}(1 - b - c) \rangle$
$\rightarrow_{49}$	$\langle \min(1, b + c), \max(0, 1 - b - c) \rangle$
$\rightarrow_{50}$	$\langle b + c - bc, 1 - b - c + bc \rangle$
$\rightarrow_{51}$	$\langle \min(\max(b, c), \max(1 - b, b), \max(c, 1 - c)),$ $\max(1 - \max(b, c), \min(1 - b, b), \min(c, 1 - c)) \rangle$
$\rightarrow_{52}$	$\langle 1 - (1 - \min(b, c))\text{sg}(1 - b - c), 1 - \min(b, c)\text{sg}(1 - b - c) \rangle$
$\rightarrow_{53}$	$\langle b + (1 - b)^2 c, (1 - b)b + (1 - b)^2(1 - c) \rangle$
$\rightarrow_{54}$	$\langle c\overline{\text{sg}}(b) + \text{sg}(b)(\overline{\text{sg}}(1 - c) + b\text{sg}(1 - c)),$ $(1 - c)\overline{\text{sg}}(b) + (1 - b)\text{sg}(b)\text{sg}(1 - c) \rangle$
$\rightarrow_{55}$	$\langle 1 - \text{sg}(1 - b - c), 1 - \overline{\text{sg}}(1 - b - c) \rangle$
$\rightarrow_{56}$	$\langle \max(\overline{\text{sg}}(1 - b), c), \min(\text{sg}(1 - b), 1 - c) \rangle$
$\rightarrow_{57}$	$\langle \max(\overline{\text{sg}}(1 - b), \text{sg}(c)), \min(\text{sg}(1 - b), \overline{\text{sg}}(c)) \rangle$
$\rightarrow_{58}$	$\langle \max(\overline{\text{sg}}(1 - b), \overline{\text{sg}}(1 - c)), 1 - \max(b, c) \rangle$
$\rightarrow_{59}$	$\langle \max(\overline{\text{sg}}(1 - b), c), 1 - \max(b, c) \rangle$
$\rightarrow_{60}$	$\langle \max(\overline{\text{sg}}(1 - b), \overline{\text{sg}}(1 - c)), \min(1 - b, \overline{\text{sg}}(c)) \rangle$
$\rightarrow_{61}$	$\langle \max(c, \min(b, d)), \min(a, d) \rangle$
$\rightarrow_{62}$	$\langle \overline{\text{sg}}(d - b), a\text{sg}(d - b) \rangle$
$\rightarrow_{63}$	$\langle 1 - (1 - b)\text{sg}(d - b), a\text{sg}(d - b) \rangle$
$\rightarrow_{64}$	$\langle c + bd, ad \rangle$
$\rightarrow_{65}$	$\langle 1 - (1 - \min(b, c))\text{sg}(d - b), \max(a, d)\text{sg}(d - b)\text{sg}(a - c) \rangle$
$\rightarrow_{66}$	$\langle c + d^2 b, bd + d^2 a \rangle$
$\rightarrow_{67}$	$\langle b\overline{\text{sg}}(1 - d) + \text{sg}(1 - d)(\overline{\text{sg}}(1 - b) + c\text{sg}(1 - b)),$ $a\overline{\text{sg}}(1 - d) + d\text{sg}(1 - d)\text{sg}(1 - b) \rangle$

(continued)

Table 1.1 (continued)

$\to_{68}$	$\langle 1 - (1-b)\mathrm{sg}(d-b), a\mathrm{sg}(d-b)\mathrm{sg}(a-c)\rangle$
$\to_{69}$	$\langle 1 - (1-b)\mathrm{sg}(d-b) - a\overline{\mathrm{sg}}(d-b)\mathrm{sg}(a-c), a\mathrm{sg}(a-c)\rangle$
$\to_{70}$	$\langle \max(\overline{\mathrm{sg}}(d), b), \min(\mathrm{sg}(d), a)\rangle$
$\to_{71}$	$\langle \max(b, c), \min(cd + d^2, a)\rangle$
$\to_{72}$	$\langle \max(b, c), \min(1 - c, a)\rangle$
$\to_{73}$	$\langle \max(1 - \max(\mathrm{sg}(d), \mathrm{sg}(1-c)), b), \min(\mathrm{sg}(1-c), a)\rangle$
$\to_{74}$	$\langle \max(\mathrm{sg}(b), \overline{\mathrm{sg}}(d)), \min(\overline{\mathrm{sg}}(b), \mathrm{sg}(d))\rangle$
$\to_{75}$	$\langle \max(c, b(a+b)), \min(d(c+d), a(b^2+a)+ab)\rangle$
$\to_{76}$	$\langle \max(c, 1-a), \min(1-c, a)\rangle$
$\to_{77}$	$\langle (1 - \min(\overline{\mathrm{sg}}(1-a), \mathrm{sg}(1-c))), \min(\overline{\mathrm{sg}}(1-a), \mathrm{sg}(1-c))\rangle$
$\to_{78}$	$\langle \max(\overline{\mathrm{sg}}(1-c), b), \min(\mathrm{sg}(d), a)\rangle$
$\to_{79}$	$\langle \max(\overline{\mathrm{sg}}(1-c), \mathrm{sg}(b)), \min(\mathrm{sg}(d), \overline{\mathrm{sg}}(1-a))\rangle$
$\to_{80}$	$\langle \max(\overline{\mathrm{sg}}(1-c), b), \min(d, a)\rangle$
$\to_{81}$	$\langle \max(\overline{\mathrm{sg}}(1-b), \overline{\mathrm{sg}}(1-c)), \min(d, \overline{\mathrm{sg}}(1-a))\rangle$
$\to_{82}$	$\langle \max(1-d, \min(d, 1-a)), \min(d, a)\rangle$
$\to_{83}$	$\langle \overline{\mathrm{sg}}(a+d-1), a\mathrm{sg}(a+d-1)\rangle$
$\to_{84}$	$\langle 1 - a\mathrm{sg}(a+d-1), a\mathrm{sg}(a+d-1)\rangle$
$\to_{85}$	$\langle 1 - d + d^2(1-a), d(1-d) + d^2\rangle$
$\to_{86}$	$\langle (1-a)\overline{\mathrm{sg}}(1-d) + \mathrm{sg}(1-d)(\overline{\mathrm{sg}}(a) + (1-d)\mathrm{sg}(d)),$ $a\overline{\mathrm{sg}}(1-d) + d\mathrm{sg}(1-d)\mathrm{sg}(a)\rangle$
$\to_{87}$	$\langle \max(\overline{\mathrm{sg}}(d), 1-a), \min(\mathrm{sg}(d), a)\rangle$
$\to_{88}$	$\langle \max(\overline{\mathrm{sg}}(d), \mathrm{sg}(1-a)), \min(\mathrm{sg}(d), \overline{\mathrm{sg}}(1-a))\rangle$
$\to_{89}$	$\langle \max(\overline{\mathrm{sg}}(d), 1-a), \min(d, a)\rangle$
$\to_{90}$	$\langle \max(\overline{\mathrm{sg}}(a), \overline{\mathrm{sg}}(d)), \min(d, \overline{\mathrm{sg}}(1-a))\rangle$
$\to_{91}$	$\langle \max(c, \min(1-c, b)), 1 - \max(b, c)\rangle$
$\to_{92}$	$\langle \overline{\mathrm{sg}}(1-b-c), \min(1-b, \mathrm{sg}(1-b-c))\rangle$
$\to_{93}$	$\langle (1 - \min(1-b, \mathrm{sg}(1-b-c)), \min(1-b, \mathrm{sg}(1-b-c))\rangle$
$\to_{94}$	$\langle c + (1-c)^2 b, (1-c)c + (1-c)^2(1-b)\rangle$
$\to_{95}$	$\langle \min(b, \overline{\mathrm{sg}}(c)) + \mathrm{sg}(c)(\overline{\mathrm{sg}}(1-b) + \min(c, \mathrm{sg}(1-b))),$ $\min(1-b, \overline{\mathrm{sg}}(c)) + \min(1-c, \mathrm{sg}(c), \mathrm{sg}(1-b))\rangle$
$\to_{96}$	$\langle \max(\overline{\mathrm{sg}}(1-c), b), \min(\mathrm{sg}(1-b), 1-c)\rangle$
$\to_{97}$	$\langle \max(\overline{\mathrm{sg}}(1-c), \mathrm{sg}(b)), \min(\mathrm{sg}(1-c), \overline{\mathrm{sg}}(b))\rangle$
$\to_{98}$	$\langle \max(\overline{\mathrm{sg}}(1-c), b), 1 - \max(b, c)\rangle$
$\to_{99}$	$\langle \max(\overline{\mathrm{sg}}(1-c), \overline{\mathrm{sg}}(1-b)), \min(1-c, \overline{\mathrm{sg}}(b))\rangle$
$\to_{100}$	$\langle \max(b\mathrm{sg}(a), c), \min(a\mathrm{sg}(b), d)\rangle$
$\to_{101}$	$\langle \max(b\mathrm{sg}(a), c\mathrm{sg}(d)), \min(a\mathrm{sg}(b), \mathrm{sg}(c)d)\rangle$
$\to_{102}$	$\langle \max(b, c\mathrm{sg}(d)), \min(a, \mathrm{sg}(c)d)\rangle$
$\to_{103}$	$\langle \max(\min(1-a, \mathrm{sg}(a)), 1-d), \min(a, \mathrm{sg}(1-a), d)\rangle$

(continued)

Table 1.1 (continued)

$\rightarrow_{104}$	$\langle\max(\min(1-a,\mathrm{sg}(a)),\min(1-d,\mathrm{sg}(d))),$ $\min(a,\mathrm{sg}(1-a),d,\mathrm{sg}(1-d))\rangle$
$\rightarrow_{105}$	$\langle\max(1-a,\min(1-d,\mathrm{sg}(d))),\min(a,d,\mathrm{sg}(1-d))\rangle$
$\rightarrow_{106}$	$\langle\max(\min(b,\mathrm{sg}(1-b)),c),\min(1-b,\mathrm{sg}(b),1-c)\rangle$
$\rightarrow_{107}$	$\langle\max(\min(b,\mathrm{sg}(1-b)),\min(c,\mathrm{sg}(1-c))),$ $\min(1-b,\mathrm{sg}(b),1-c,\mathrm{sg}(c))\rangle$
$\rightarrow_{108}$	$\langle\max(b,\min(c,\mathrm{sg}(1-c))),\min(1-b,1-c,\mathrm{sg}(c))\rangle$
$\rightarrow_{109}$	$\langle b+\min(\overline{\mathrm{sg}}(1-a),c),ab+\min(\overline{\mathrm{sg}}(1-a),d))\rangle$
$\rightarrow_{110}$	$\langle\max(b,c),\min(ab+\overline{\mathrm{sg}}(1-a),d)\rangle$
$\rightarrow_{111}$	$\langle\max(b,cd+\overline{\mathrm{sg}}(1-c)),\min(ab+\overline{\mathrm{sg}}(1-a),$ $d(cd+\overline{\mathrm{sg}}(1-c))+\overline{\mathrm{sg}}(1-d))\rangle$
$\rightarrow_{112}$	$\langle b+c-bc,ab+\overline{\mathrm{sg}}(1-a)d\rangle$
$\rightarrow_{113}$	$\langle b+cd-b(cd+\overline{\mathrm{sg}}(1-c)),$ $(ab+\overline{\mathrm{sg}}(1-a))(d(cd+\overline{\mathrm{sg}}(1-c))+\overline{\mathrm{sg}}(1-d))\rangle$
$\rightarrow_{114}$	$\langle 1-a+\min(\overline{\mathrm{sg}}(1-a),1-d),a(1-a)+\min(\overline{\mathrm{sg}}(1-a),d)\rangle$
$\rightarrow_{115}$	$\langle 1-\min(a,d),\min(a(1-a)+\overline{\mathrm{sg}}(1-a),d)\rangle$
$\rightarrow_{116}$	$\langle\max(1-a,(1-d)d+\overline{\mathrm{sg}}(d)),$ $\min(a(1-a)+\overline{\mathrm{sg}}(1-a),d((1-d)d+\overline{\mathrm{sg}}(d))+\overline{\mathrm{sg}}(1-d))\rangle$
$\rightarrow_{117}$	$\langle 1-a-d+ad,(a(1-a)+\overline{\mathrm{sg}}(1-a))d\rangle$
$\rightarrow_{118}$	$\langle 1-a+(1-d)d-(1-a)((1-d)d+\overline{\mathrm{sg}}(d)),$ $(a(1-a)+\overline{\mathrm{sg}}(1-a))d((1-d)d+\overline{\mathrm{sg}}(d))+\overline{\mathrm{sg}}(1-d)\rangle$
$\rightarrow_{119}$	$\langle b+\min(\overline{\mathrm{sg}}(b),c),(1-b)b+\min(\overline{\mathrm{sg}}(b),1-c)\rangle$
$\rightarrow_{120}$	$\langle\max(b,c),\min((1-b)b+\overline{\mathrm{sg}}(b),1-c)\rangle$
$\rightarrow_{121}$	$\langle\max(b,c(1-c)+\overline{\mathrm{sg}}(1-c)),$ $\min((1-b)b+\overline{\mathrm{sg}}(b),(1-c)(c(1-c)+\overline{\mathrm{sg}}(1-c)))+\overline{\mathrm{sg}}(c)\rangle$
$\rightarrow_{122}$	$\langle b+c-bc,((1-b)b+\overline{\mathrm{sg}}(b))(1-c)\rangle$
$\rightarrow_{123}$	$\langle b+c(1-c)-(b(c(1-c)+\overline{\mathrm{sg}}(1-c))),$ $((1-b)b+\overline{\mathrm{sg}}(b))((1-c)(c(1-c)+\overline{\mathrm{sg}}(1-c)))+\overline{\mathrm{sg}}(c))\rangle$
$\rightarrow_{124}$	$\langle c+\min(\overline{\mathrm{sg}}(1-d),b),cd+\min(\overline{\mathrm{sg}}(1-d),a)\rangle$
$\rightarrow_{125}$	$\langle\max(b,c),\min(cd+\overline{\mathrm{sg}}(1-d),a)\rangle$
$\rightarrow_{126}$	$\langle\max(c,ab+\overline{\mathrm{sg}}(1-b)),$ $\min(cd+\overline{\mathrm{sg}}(1-d),a(ab+\overline{\mathrm{sg}}(1-b))+\overline{\mathrm{sg}}(1-a))\rangle$
$\rightarrow_{127}$	$\langle b+c-bc,(cd+\overline{\mathrm{sg}}(1-d))a\rangle$
$\rightarrow_{128}$	$\langle c+ab-c(ab+\overline{\mathrm{sg}}(1-b)),$ $(cd+\overline{\mathrm{sg}}(1-d))(a(ab+\overline{\mathrm{sg}}(1-b))+\overline{\mathrm{sg}}(1-a))\rangle$
$\rightarrow_{129}$	$\langle 1-d+\min(\overline{\mathrm{sg}}(1-d),1-a),d(1-d)+\min(\overline{\mathrm{sg}}(1-d),a)\rangle$
$\rightarrow_{130}$	$\langle 1-\min(d,a),\min(d(1-d)+\overline{\mathrm{sg}}(1-d),a)\rangle$
$\rightarrow_{131}$	$\langle\max(1-d,(1-a)a+\overline{\mathrm{sg}}(a)),$ $\min(d(1-d)+\overline{\mathrm{sg}}(1-d),a((1-a)a+\overline{\mathrm{sg}}(a))+\overline{\mathrm{sg}}(1-a))\rangle$
$\rightarrow_{132}$	$\langle 1-ad,(d(1-d)+\overline{\mathrm{sg}}(1-d))a\rangle$

(continued)

Table 1.1 (continued)

$\rightarrow_{133}$	$\langle 1 - d + (1 - a)a - (1 - d)((1 - a)a + \overline{sg}(a)),$ $(d(1 - d) + \overline{sg}(1 - d))(a((1 - a)a + \overline{sg}(a)) + \overline{sg}(1 - a)))\rangle$
$\rightarrow_{134}$	$\langle c + \min(\overline{sg}(c), b), (1 - c)c + \min(\overline{sg}(c), (1 - b))\rangle$
$\rightarrow_{135}$	$\langle \max(b, c), \min((1 - c)c + \overline{sg}(c), 1 - b)\rangle$
$\rightarrow_{136}$	$\langle \max(c, b(1 - b) + \overline{sg}(1 - b)),$ $\min((1 - c)c + \overline{sg}(c), (1 - b)(b(1 - b) + \overline{sg}(1 - b)) + \overline{sg}(b)))\rangle$
$\rightarrow_{137}$	$\langle b + c - bc, ((1 - c)c + \overline{sg}(c))(1 - b)\rangle$
$\rightarrow_{138}$	$\langle c + b(1 - b) - c(b(1 - b) + \overline{sg}(1 - b)),$ $((1 - c)c + \overline{sg}(c))((1 - b)(b(1 - b) + \overline{sg}(1 - b)) + \overline{sg}(b)))\rangle$
$\rightarrow_{139}$	$\langle \frac{b+c}{2}, \frac{a+d}{2} \rangle$
$\rightarrow_{140}$	$\langle \frac{b+c+\min(b,c)}{3}, \frac{a+d+\max(a,d)}{3} \rangle$
$\rightarrow_{141}$	$\langle \frac{b+c+\max(b,c)}{3}, \frac{a+d+\min(a,d)}{3} \rangle$
$\rightarrow_{142}$	$\langle \frac{3-a-d-\max(a,d)}{3}, \frac{a+d+\max(a,d)}{3} \rangle$
$\rightarrow_{143}$	$\langle \frac{1-a+c+\min(1-a,c)}{3}, \frac{2+a-c-\min(1-a,c)}{3} \rangle$
$\rightarrow_{144}$	$\langle \frac{1+b-d+\min(b,1-d)}{3}, \frac{2-b+d-\min(b,1-d)}{3} \rangle$
$\rightarrow_{145}$	$\langle \frac{b+c+\min(b,c)}{3}, \frac{3-b-c-\min(b,c)}{3} \rangle$
$\rightarrow_{146}$	$\langle \frac{3-a-d-\min(a,d)}{3}, \frac{a+d+\min(a,d)}{3} \rangle$
$\rightarrow_{147}$	$\langle \frac{1-a+c+\max(1-a,c)}{3}, \frac{2+a-c-\max(1-a,c)}{3} \rangle$
$\rightarrow_{148}$	$\langle \frac{1+b-d+\max(b,1-d)}{3}, \frac{2-b+d-\max(b,1-d)}{3} \rangle$
$\rightarrow_{149}$	$\langle \frac{b+c+\max(b,c)}{3}, \frac{3-b-c-\max(b,c)}{3} $
$\rightarrow_{150,\lambda}$	$\langle \frac{b+c+\lambda-1}{2\lambda}, \frac{a+d+\lambda-1}{2\lambda},$ where $\lambda \geq 1$
$\rightarrow_{151,\gamma}$	$\langle \frac{b+c+\gamma}{2\gamma+1}, \frac{a+d+\gamma-1}{2\gamma+1},$ where $\gamma \geq 1$
$\rightarrow_{152,\alpha,\beta}$	$\langle \frac{b+c+\alpha-1}{\alpha+\beta}, \frac{a+d+\beta-1}{\alpha+\beta}$ where $\alpha \geq 1, \beta \in [1, \alpha]$
$\rightarrow_{153,\varepsilon,\eta}$	$\langle \min(1, \max(c, b + \varepsilon)), \max(0, \min(d, a - \eta))\rangle$ where $\varepsilon, \eta \in [0, 1]$ and $\varepsilon \leq \eta < 1$
$\rightarrow_{154,\lambda}$	$\langle \frac{-a+c+\lambda}{2\lambda}, \frac{a-c+\lambda}{2\lambda} \rangle,$ where $\lambda \geq 1$
$\rightarrow_{155,\lambda}$	$\langle \frac{1-a-d+\lambda}{2\lambda}, \frac{a+d+\lambda-1}{2\lambda} \rangle,$ where $\lambda \geq 1$

(continued)

Table 1.1 (continued)

$\rightarrow_{156,\lambda}$	$\langle\frac{b+c+\lambda-1}{2\lambda}, \frac{1-b-c+\lambda}{2\lambda}\rangle$, where $\lambda \geq 1$
$\rightarrow_{157,\lambda}$	$\langle\frac{b-d+\lambda}{2\lambda}, \frac{-b+d+\lambda}{2\lambda}\rangle$, where $\lambda \geq 1$
$\rightarrow_{158,\gamma}$	$\langle\frac{1-a+c+\gamma}{2\gamma+1}, \frac{a-c+\gamma}{2\gamma+1}\rangle$, where $\gamma \geq 1$
$\rightarrow_{159,\gamma}$	$\langle\frac{2-a-d+\gamma}{2\gamma+1}, \frac{a+d+\gamma-1}{2\gamma+1}\rangle$, where $\gamma \geq 1$
$\rightarrow_{160,\gamma}$	$\langle\frac{b-d+\gamma+1}{2\gamma+1}, \frac{-b+d+\gamma}{2\gamma+1}\rangle$, where $\gamma \geq 1$
$\rightarrow_{161,\gamma}$	$\langle\frac{b+c+\gamma}{2\gamma+1}, \frac{1-b-c+\gamma}{2\gamma+1}\rangle$, where $\gamma \geq 1$
$\rightarrow_{162,\alpha,\beta}$	$\langle\frac{-a+c+\alpha}{\alpha+\beta}, \frac{a-c+\beta}{\alpha+\beta}\rangle$, where $\alpha \geq 1$, $\beta \in [1, \alpha]$
$\rightarrow_{163,\alpha,\beta}$	$\langle\frac{1-a-d+\alpha}{\alpha+\beta}, \frac{a+d+\beta-1}{\alpha+\beta}\rangle$, where $\alpha \geq 1$, $\beta \in [1, \alpha]$
$\rightarrow_{164,\alpha,\beta}$	$\langle\frac{b-d+\alpha}{\alpha+\beta}, \frac{-b+d+\beta}{\alpha+\beta}\rangle$, where $\alpha \geq 1$, $\beta \in [1, \alpha]$
$\rightarrow_{165,\alpha,\beta}$	$\langle\frac{b+c+\alpha-1}{\alpha+\beta}, \frac{1-b-c+\beta}{\alpha+\beta}\rangle$, where $\alpha \geq 1$, $\beta \in [1, \alpha]$
$\rightarrow_{166}$	$\langle\max(b, \min(a, c)), \min(a, \max(b, d))\rangle$
$\rightarrow_{167}$	$\langle\max(1-a, \min(a, c)), \min(a, 1-\min(a, c))\rangle$
$\rightarrow_{168}$	$\langle\max(1-a, \min(a, 1-d)), 1-\max(1-a, \min(a, 1-d))\rangle$
$\rightarrow_{169}$	$\langle\max(b, \min(1-b, c)), 1-\max(b, \min(1-b, c))\rangle$
$\rightarrow_{170}$	$\langle\max(b, \min(1-b, 1-d)), 1-\max(b, \min(1-b, 1-d))\rangle$
$\rightarrow_{171}$	$\langle\overline{sg}(\max(a, d) - \max(b, c)), sg(\max(a, d) - \max(b, c))\rangle$
$\rightarrow_{172}$	$\langle\overline{sg}(a - c), sg(a - c)\rangle$
$\rightarrow_{173}$	$\langle\overline{sg}(a + d - 1), sg(a + d - 1)\rangle$
$\rightarrow_{174}$	$\langle\overline{sg}(1 - b - c), sg(1 - b - c)\rangle$
$\rightarrow_{175}$	$\langle\overline{sg}(d - b), sg(d - b)\rangle$
$\rightarrow_{176}$	$\langle\overline{sg}(a - c) + sg(a - c)\max(b, c), sg(a - c)\min(a, d)\rangle$
$\rightarrow_{177}$	$\langle\overline{sg}(a - c) + sg(a - c)\max(1 - a, c), sg(a - c)\min(a, 1 - c)\rangle$
$\rightarrow_{178}$	$\langle\overline{sg}(a - 1 + d) + sg(a - 1 + d)(1 - \min(a, d)),$ $sg(a - 1 + d)\min(a, d)\rangle$
$\rightarrow_{179}$	$\langle\overline{sg}(1 - b - c) + sg(1 - b - c)\max(b, c),$ $sg(1 - b - c)(1 - \max(b, c))\rangle$
$\rightarrow_{180}$	$\langle\overline{sg}(d - b) + sg(d - b)\max(b, 1 - d), sg(d - b)\min(1 - b, d)\rangle$
$\rightarrow_{181}$	$\langle1 - sg(a).(1 - c), d.sg(a)\rangle$
$\rightarrow_{182}$	$\langle1 - sg(a).(1 - c), (1 - c).sg(a)\rangle$
$\rightarrow_{183}$	$\langle1 - sg(a).d, d.sg(a)\rangle$
$\rightarrow_{184}$	$\langle1 - sg(1 - b).d, d.sg(1 - b)\rangle$
$\rightarrow_{185}$	$\langle1 - sg(1 - b).(1 - c), (1 - c).sg(1 - b)\rangle$

$$\langle 0, 1 \rangle \rightarrow_i \langle 0, 1 \rangle = \begin{cases} \langle 1, 0 \rangle, & \text{for } i = 1, \dots, 99, 102, 105, 108, \dots, 127, \\ & 129, \dots, 132, 134, \dots, 137, 153, 166, \dots, 185 \\ \langle 0, 0 \rangle, & \text{for } i = 100, 101, 103, 104, 106, 107, 128, \\ & 133, 138 \\ \langle \frac{1}{2}, \frac{1}{2} \rangle, & \text{for } i = 139 \\ \langle \frac{1}{3}, \frac{2}{3} \rangle, & \text{for } i = 140, 142, \dots, 145 \\ \langle \frac{2}{3}, \frac{1}{3} \rangle, & \text{for } i = 141, 146, \dots, 149 \\ \langle \frac{\lambda+1}{2\lambda}, \frac{\lambda-1}{2\lambda} \rangle, & \text{for } i = 150, 154, \dots, 157 \\ \langle \frac{\gamma+2}{2\gamma+1}, \frac{\gamma-1}{2\gamma+1} \rangle, & \text{for } i = 151, 158, \dots, 161 \\ \langle \frac{\alpha+1}{\alpha+\beta}, \frac{\beta-1}{\alpha+\beta} \rangle, & \text{for } i = 152, 162, \dots, 165. \end{cases}$$

$$\langle 0, 1 \rangle \rightarrow_i \langle 1, 0 \rangle = \begin{cases} \langle 1, 0 \rangle, & \text{for } i = 1, \dots, 100, 102, 103, 105, 106, 108, \\ & \dots, 112, 114, \dots, 117, \dots, 121, 123, \dots, 126, \\ & 128, \dots, 132, 134, \dots, 137, 139, \dots, 149, 153, \\ & 166, \dots, 185 \\ \langle 0, 0 \rangle, & \text{for } i = 101, 104, 107, 113, 118, 122, 127, \\ & 133, 138 \\ \langle \frac{\lambda+1}{2\lambda}, \frac{\lambda-1}{2\lambda} \rangle, & \text{for } i = 150, 154, \dots, 157 \\ \langle \frac{\gamma+2}{2\gamma+1}, \frac{\gamma-1}{2\gamma+1} \rangle, & \text{for } i = 151, 158, \dots, 161 \\ \langle \frac{\alpha+1}{\alpha+\beta}, \frac{\beta-1}{\alpha+\beta} \rangle, & \text{for } i = 152, 162, \dots, 165. \end{cases}$$

$$\langle 1, 0 \rangle \rightarrow_i \langle 0, 1 \rangle = \begin{cases} \langle 0, 1 \rangle, & \text{for } i = 1, \dots, 99, 109, \dots, 149, 166, \dots, 168, \\ & 170, \dots, 185 \\ \langle 0, 0 \rangle, & \text{for } i = 100, \dots, 108, 169 \\ \langle \frac{\lambda-1}{2\lambda}, \frac{\lambda+1}{2\lambda} \rangle, & \text{for } i = 150, 154, \dots, 157 \\ \langle \frac{\gamma}{2\gamma+1}, \frac{\gamma+1}{2\gamma+1} \rangle, & \text{for } i = 151, 158, \dots, 161 \\ \langle \frac{\alpha-1}{\alpha+\beta}, \frac{\beta+1}{\alpha+\beta} \rangle, & \text{for } i = 152, 162, \dots, 165 \\ \langle \varepsilon, 1 - \eta \rangle, & \text{for } i = 153. \end{cases}$$

$$\langle 1,0\rangle \rightarrow_i \langle 1,0\rangle = \begin{cases} \langle 1,0\rangle, & \text{for } i = 1,\ldots,100, 103, 106, 109,\ldots,112, \\ & 114,\ldots,117, 119,\ldots,122, 124,\ldots,138, \\ & 153, 166,\ldots,185 \\ \langle 0,0\rangle, & \text{for } i = 101, 102, 104, 105, 107, 108, 113, 123 \\ \langle 0,\frac{1}{2}\rangle, & \text{for } i = 139, 150, 155,\ldots,157 \\ \langle \frac{1}{3},\frac{2}{3}\rangle, & \text{for } i = 140, 142,\ldots,145 \\ \langle \frac{2}{3},\frac{1}{3}\rangle, & \text{for } i = 141, 146,\ldots,149 \\ \langle \frac{\gamma+1}{2\gamma+1},\frac{\gamma}{2\gamma+1}\rangle, & \text{for } i = 151, 158,\ldots,161 \\ \langle \frac{\alpha}{\alpha+\beta},\frac{\beta}{\alpha+\beta}\rangle, & \text{for } i = 152, 162,\ldots,165. \end{cases}$$

It is interesting to mention that the well-known axiom of the classical logic $A \rightarrow A$ is an IFT for implications $\rightarrow_1, \ldots, \rightarrow_9, \rightarrow_{11}, \ldots, \rightarrow_{15}, \rightarrow_{17}, \rightarrow_{18}, \rightarrow_{20}, \ldots, \rightarrow_{24}, \rightarrow_{27}, \ldots, \rightarrow_{38}, \rightarrow_{40}, \rightarrow_{42}, \rightarrow_{44}, \ldots, \rightarrow_{53}, \rightarrow_{55}, \rightarrow_{57}, \rightarrow_{59}, \ldots, \rightarrow_{66}, \rightarrow_{68}, \rightarrow_{69}, \rightarrow_{71}, \rightarrow_{72}, \rightarrow_{74}, \ldots, \rightarrow_{77}, \rightarrow_{79}, \ldots, \rightarrow_{85}, \rightarrow_{88}, \ldots, \rightarrow_{94}, \rightarrow_{97}, \ldots, \rightarrow_{139}, \rightarrow_{141}, \rightarrow_{146}, \ldots, \rightarrow_{170}, \rightarrow_{176}, \ldots, \rightarrow_{185}$, while it is just a tautology for implications $\rightarrow_2, \rightarrow_3, \rightarrow_5, \rightarrow_8, \rightarrow_{11}, \rightarrow_{14}, \rightarrow_{15}, \rightarrow_{20}, \rightarrow_{23}, \rightarrow_{24}, \rightarrow_{27}, \rightarrow_{31}, \rightarrow_{32}, \rightarrow_{34}, \rightarrow_{37}, \rightarrow_{40}, \rightarrow_{42}, \rightarrow_{47}, \ldots, \rightarrow_{49}, \rightarrow_{52}, \rightarrow_{55}, \rightarrow_{57}, \rightarrow_{62}, \rightarrow_{63}, \rightarrow_{65}, \rightarrow_{68}, \rightarrow_{69}, \rightarrow_{74}, \rightarrow_{77}, \rightarrow_{79}, \rightarrow_{83}, \rightarrow_{84}, \rightarrow_{88}, \rightarrow_{92}, \rightarrow_{93}, \rightarrow_{97}, \rightarrow_{176}, \ldots, \rightarrow_{185}.$

The intuitionistic fuzzy implications that satisfy the following equalities

$$\langle 0,1\rangle \rightarrow_i \langle 0,1\rangle = \langle 1,0\rangle,$$

$$\langle 0,1\rangle \rightarrow_i \langle 1,0\rangle = \langle 1,0\rangle,$$

$$\langle 1,0\rangle \rightarrow_i \langle 0,1\rangle = \langle 0,1\rangle,$$

$$\langle 1,0\rangle \rightarrow_i \langle 1,0\rangle = \langle 1,0\rangle,$$

as standard tautologies, will be called "implications of (fully) tautological type" (T-implications), while the implications that satisfy these equalities as IFTs, will be called "implications from IFT type" (I-implications), and the rest are incorrect and will be denoted as N-implications.

1.4 Intuitionistic Fuzzy Negations

The currently existing intuitionistic fuzzy implications generate the intuitionistic fuzzy negations [57, 59–62] using the formula

$$\neg\langle a,b\rangle = \langle a,b\rangle \rightarrow \langle 0,1\rangle,$$

where $\rightarrow$ is any of the defined implications. Paper [62] is written together with D. Dimitrov and paper [60] – with N. Angelova. The full list of different negations is shown in Table 1.2. As it can be seen, a lot of different implications generate exactly one negation.

The relationships between the negations and implications are shown in Table 1.3.

1.5 Properties of Intuitionistic Fuzzy Implications and Negations

Here, we study some interesting properties of the intuitionistic fuzzy implications and negations. A part of them are published in [63].

For the well-known formula

$$(A \to B) \vee (B \to A), \tag{1.5.1}$$

in [46] the following two theorems are proved. Here, they are extended with check of the properties of the implications $\to_{153}, \ldots, \to_{185}$ (given after the symbol $*$).

Theorem 1.5.1 *For every two formulas A and B, (1.5.1) is an IFT for the intuitionistic fuzzy implications* $\to_1, \ldots, \to_6, \to_8, \to_9, \to_{11}, \to_{13}, \to_{14}, \to_{17}, \to_{18}, \to_{20}, \ldots, \to_{24}, \to_{27}, \ldots, \to_{38}, \to_{40}, \to_{42}, \to_{44}, \to_{45}, \to_{61}, \ldots, \to_{66}, \to_{68}, \to_{69}, \to_{71}, \to_{72}, \to_{74}, \ldots, \to_{77}, \to_{79}, \ldots, \to_{85}, \to_{88}, \ldots, \to_{90}, \to_{100}, \ldots, \to_{105}, \to_{109}, \ldots, \to_{118}, \to_{124}, \ldots, \to_{133}, \overset{*}{\to}_{139}, \to_{141}, \to_{146}, \ldots, \to_{148}, \to_{150}, \ldots, \to_{155}, \to_{157}, \ldots, \to_{160}, \to_{162}, \ldots, \to_{164}, \to_{166}, \ldots, \to_{170}, \to_{176}, \ldots, \to_{178}, \to_{180}, \ldots, \to_{183}, \to_{185}.$

Theorem 1.5.2 *For every two formulas A and B, (1.5.1) is a tautology for the intuitionistic fuzzy implications* $\to_2, \to_3, \to_8, \to_{11}, \to_{20}, \to_{23}, \to_{31}, \to_{32}, \to_{34}, \to_{37}, \to_{40}, \to_{42}, \to_{62}, \to_{63}, \to_{65}, \to_{68}, \to_{74}, \to_{77}, \to_{83}, \to_{84}, \to_{88}, \overset{*}{\to}_{153}, \to_{176}, \ldots, \to_{178}, \to_{180}, \ldots, \to_{183}, \to_{185}.$

G.F. Rose's formula [64, 65] has the form:

$$((\neg\neg A \to A) \to (\neg\neg A \vee \neg A)) \to (\neg\neg A \vee \neg A). \tag{1.5.2}$$

For it, the following two theorems are valid.

Theorem 1.5.3 *For each formula A, (1.5.2) is an IFT for the intuitionistic fuzzy implications* $\to_1, \ldots, \to_9, \to_{11}, \ldots, \to_{38}, \to_{40}, \ldots, \to_{53}, \to_{55}, \ldots, \to_{57}, \to_{61}, \to_{62}, \to_{64}, \ldots, \to_{67}, \to_{71}, \to_{72}, \to_{74}, \ldots, \to_{77}, \to_{79}, \ldots, \to_{83}, \to_{85}, \to_{86}, \to_{88}, \ldots, \to_{91}, \to_{94}, \to_{95}, \to_{97}, \ldots, \to_{107}, \to_{109}, \ldots, \to_{137}, \to_{151}, \to_{153}, \to_{158}, \ldots, \to_{161}, \to_{166}, \ldots, \to_{185}.$

Theorem 1.5.4 *For each formula A, (1.5.2) is a tautology for the intuitionistic fuzzy implications* $\to_2, \to_3, \to_8, \to_{11}, \to_{14}, \ldots, \to_{16}, \to_{19}, \to_{20}, \to_{23}, \to_{24}, \to_{31}, \to_{32}, \to_{37}, \to_{40}, \ldots, \to_{45}, \to_{47}, \to_{48}, \to_{55}, \ldots, \to_{57}, \to_{62}, \to_{65}, \to_{74}, \to_{77}, \to_{83}, \to_{88}, \to_{90}, \to_{97}, \to_{99}, \to_{153}, \to_{171}, \to_{180}.$

Table 1.2 List of the intuitionistic fuzzy negations

$\neg_1$	$\langle b, a \rangle$
$\neg_2$	$\langle \overline{sg}(a), sg(a) \rangle$
$\neg_3$	$\langle b, a.b + a^2 \rangle$
$\neg_4$	$\langle b, 1 - b \rangle$
$\neg_5$	$\langle \overline{sg}(1 - b), sg(1 - b) \rangle$
$\neg_6$	$\langle \overline{sg}(1 - b), sg(a) \rangle$
$\neg_7$	$\langle \overline{sg}(1 - b), a \rangle$
$\neg_8$	$\langle 1 - a, a \rangle$
$\neg_9$	$\langle \overline{sg}(a), a \rangle$
$\neg_{10}$	$\langle \overline{sg}(1 - b), 1 - b \rangle$
$\neg_{11}$	$\langle sg(b), \overline{sg}(b) \rangle$
$\neg_{12}$	$\langle b.(b + a), a.(b^2 + a + b.a) \rangle$
$\neg_{13}$	$\langle sg(1 - a), \overline{sg}(1 - a) \rangle$
$\neg_{14}$	$\langle sg(b), \overline{sg}(1 - a) \rangle$
$\neg_{15}$	$\langle \overline{sg}(1 - b), \overline{sg}(1 - a) \rangle$
$\neg_{16}$	$\langle \overline{sg}(a), \overline{sg}(1 - a) \rangle$
$\neg_{17}$	$\langle \overline{sg}(1 - b), \overline{sg}(b) \rangle$
$\neg_{18}$	$\langle b.sg(a), a.sg(b) \rangle$
$\neg_{19}$	$\langle b.sg(a), 0 \rangle$
$\neg_{20}$	$\langle b, 0 \rangle$
$\neg_{21}$	$\langle \min(1 - a, sg(a)), \min(a, sg(1 - a)) \rangle$
$\neg_{22}$	$\langle \min(1 - a, sg(a)), 0 \rangle$
$\neg_{23}$	$\langle 1 - a, 0 \rangle$
$\neg_{24}$	$\langle \min(b, sg(1 - b)), \min(1 - b, sg(b)) \rangle$
$\neg_{25}$	$\langle \min(b, sg(1 - b)), 0 \rangle$
$\neg_{26}$	$\langle b, a.b + \overline{sg}(1 - a) \rangle$
$\neg_{27}$	$\langle 1 - a, a.(1 - a) + \overline{sg}(1 - a) \rangle$
$\neg_{28}$	$\langle b, (1 - b).b + \overline{sg}(b) \rangle$
$\neg_{29}$	$\langle \max(0, b.a + \overline{sg}(1 - b)), \min(1, a.(b.a + \overline{sg}(1 - b)) + \overline{sg}(1 - a)) \rangle$
$\neg_{30}$	$\langle a.b, \ a.(a.b + \overline{sg}(1 - b)) + \overline{sg}(1 - a) \rangle$
$\neg_{31}$	$\langle \max(0, (1 - a).a + \overline{sg}(a)), \min(1, a.((1 - a).a + \overline{sg}(a)) + \overline{sg}(1 - a)) \rangle$
$\neg_{32}$	$\langle (1 - a).a, \ a.((1 - a).a + \overline{sg}(a)) + \overline{sg}(1 - a) \rangle$
$\neg_{33}$	$\langle b.(1 - b) + \overline{sg}(1 - b), (1 - b).(b.(1 - b) + \overline{sg}(1 - b)) + \overline{sg}(b)) \rangle$
$\neg_{34}$	$\langle b.(1 - b), \ (1 - b).(b.(1 - b) + \overline{sg}(1 - b)) + \overline{sg}(b) \rangle$
$\neg_{35}$	$\langle \frac{b}{2}, \frac{1+a}{2} \rangle$
$\neg_{36}$	$\langle \frac{b}{3}, \frac{2+a}{3} \rangle$
$\neg_{37}$	$\langle \frac{2b}{3}, \frac{2a+1}{3} \rangle$
$\neg_{38}$	$\langle \frac{1-a}{3}, \frac{2+a}{3} \rangle$

(continued)

Table 1.2 (continued)

$\neg_{39}$	$\langle \frac{b}{3}, \frac{3-b}{3} \rangle$
$\neg_{40}$	$\langle \frac{2-2a}{3}, \frac{1+2a}{3} \rangle$
$\neg_{41}$	$\langle \frac{2b}{3}, \frac{3-2b}{3} \rangle$
$\neg_{42,\lambda}$	$\langle \frac{b+\lambda-1}{2\lambda}, \frac{a+\lambda}{2\lambda} \rangle$, where $\lambda \geq 1$
$\neg_{43,\gamma}$	$\langle \frac{b+\gamma}{2\gamma+1}, \frac{a+\gamma}{2\gamma+1} \rangle$, where $\gamma \geq 1$
$\neg_{44,\alpha,\beta}$	$\langle \frac{b+\alpha-1}{\alpha+\beta}, \frac{a+\beta}{\alpha+\beta} \rangle$, where $\alpha \geq 1$, $\beta \in [1, \alpha]$
$\neg_{45,\varepsilon,\eta}$	$\langle \min(1, b + \varepsilon), \max(0, a - \eta) \rangle$, where $\varepsilon, \eta \in [0, 1]$ and $\varepsilon \leq \eta < 1$
$\neg_{46,\lambda}$	$\langle \frac{\lambda-a}{2\lambda}, \frac{a+\lambda}{2\lambda} \rangle$, where $\lambda \geq 1$
$\neg_{47,\lambda}$	$\langle \frac{b+\lambda-1}{2\lambda}, \frac{1-b+\lambda}{2\lambda} \rangle$, where $\lambda \geq 1$
$\neg_{48,\gamma}$	$\langle \frac{1-a+\gamma}{2\gamma+1}, \frac{a+\gamma}{2\gamma+1} \rangle$, where $\gamma \geq 1$
$\neg_{49,\gamma}$	$\langle \frac{b+\gamma}{2\gamma+1}, \frac{1-b+\gamma}{2\gamma+1} \rangle$, where $\gamma \geq 1$
$\neg_{50,\alpha,\beta}$	$\langle \frac{b-1+\alpha}{\alpha+\beta}, \frac{a+\beta}{\alpha+\beta} \rangle$, where $\alpha \geq 1$, $\beta \in [1, \alpha]$
$\neg_{51,\alpha,\beta}$	$\langle \frac{b-1+\alpha}{\alpha+\beta}, \frac{1-b+\beta}{\alpha+\beta} \rangle$, where $\alpha \geq 1$, $\beta \in [1, \alpha]$
$\neg_{52}$	$\langle 1 - a, \min(1, 1 - a) \rangle$
$\neg_{53}$	$\langle \overline{sg}(a) + sg(a)b, a \rangle$

Now, we discuss the following new formulas, inspired by (1.5.2):

$$(A \vee \neg A) \rightarrow (A \rightarrow \neg\neg A) \tag{1.5.3}$$

$$(\neg\neg A \vee \neg A) \rightarrow (A \rightarrow \neg\neg A) \tag{1.5.4}$$

$$(A \rightarrow \neg\neg A) \rightarrow (A \vee \neg A) \tag{1.5.5}$$

$$(A \rightarrow \neg\neg A) \rightarrow (\neg\neg A \vee \neg A) \tag{1.5.6}$$

Obviously, in the classical propositional calculus, all these four formulas are tautologies. Now, we study their properties in the intuitionistic fuzzy case.

First, we mention that there are intuitionistic fuzzy implications for which

$$V((A \vee \neg A) \rightarrow (A \rightarrow \neg\neg A)) = V((\neg\neg A \vee \neg A) \rightarrow (A \rightarrow \neg\neg A)) \tag{1.5.7}$$

and others, for which (1.5.7) is not valid.

For example, we see that

$$V((A \vee \neg_8 A) \rightarrow_{34} (A \rightarrow_{34} \neg_8\neg_8 A))$$

Table 1.3 Relationships between negations and implications

$\neg 1$	$\rightarrow 1, \rightarrow 4, \rightarrow 5, \rightarrow 6, \rightarrow 7, \rightarrow 10, \rightarrow 13, \rightarrow 61, \rightarrow 63, \rightarrow 64, \rightarrow 66, \rightarrow 67, \rightarrow 68,$ $\rightarrow 69, \rightarrow 70, \rightarrow 71, \rightarrow 72, \rightarrow 73, \rightarrow 78, \rightarrow 80, \rightarrow 124, \rightarrow 125, \rightarrow 127, \rightarrow 166$
$\neg 2$	$\rightarrow 2, \rightarrow 3, \rightarrow 8, \rightarrow 11, \rightarrow 16, \rightarrow 20, \rightarrow 31, \rightarrow 32, \rightarrow 37, \rightarrow 40, \rightarrow 41, \rightarrow 42$ $\rightarrow 172, \rightarrow 173, \rightarrow 181, \rightarrow 182, \rightarrow 183$
$\neg 3$	$\rightarrow 9, \rightarrow 17, \rightarrow 21$
$\neg 4$	$\rightarrow 12, \rightarrow 18, \rightarrow 22, \rightarrow 46, \rightarrow 49, \rightarrow 50, \rightarrow 51, \rightarrow 53, \rightarrow 54, \rightarrow 91, \rightarrow 93, \rightarrow 94,$ $\rightarrow 95, \rightarrow 96, \rightarrow 98, \rightarrow 134, \rightarrow 135, \rightarrow 137, \rightarrow 169, \rightarrow 170, \rightarrow 179, \rightarrow 180$
$\neg 5$	$\rightarrow 14, \rightarrow 15, \rightarrow 19, \rightarrow 23, \rightarrow 47, \rightarrow 48, \rightarrow 52, \rightarrow 55, \rightarrow 56, \rightarrow 57, \rightarrow 171, \rightarrow 174,$ $\rightarrow 175, \rightarrow 184, \rightarrow 185$
$\neg 6$	$\rightarrow 24, \rightarrow 26, \rightarrow 27, \rightarrow 65$
$\neg 7$	$\rightarrow 25, \rightarrow 28, \rightarrow 29, \rightarrow 62$
$\neg 8$	$\rightarrow 30, \rightarrow 33, \rightarrow 34, \rightarrow 35, \rightarrow 36, \rightarrow 38, \rightarrow 39, \rightarrow 76, \rightarrow 82, \rightarrow 84, \rightarrow 85, \rightarrow 86,$ $\rightarrow 87, \rightarrow 89, \rightarrow 129, \rightarrow 130, \rightarrow 132, \rightarrow 167, \rightarrow 168, \rightarrow 177, \rightarrow 178$
$\neg 9$	$\rightarrow 43, \rightarrow 44, \rightarrow 45, \rightarrow 83$
$\neg 10$	$\rightarrow 58, \rightarrow 59, \rightarrow 60, \rightarrow 92$
$\neg 11$	$\rightarrow 74, \rightarrow 97$
$\neg 12$	$\rightarrow 75$
$\neg 13$	$\rightarrow 77, \rightarrow 88$
$\neg 14$	$\rightarrow 79$
$\neg 15$	$\rightarrow 81$
$\neg 16$	$\rightarrow 90$
$\neg 17$	$\rightarrow 99$
$\neg 18$	$\rightarrow 100$
$\neg 19$	$\rightarrow 101$
$\neg 20$	$\rightarrow 102, \rightarrow 108$
$\neg 21$	$\rightarrow 103$
$\neg 22$	$\rightarrow 104$
$\neg 23$	$\rightarrow 105$
$\neg 24$	$\rightarrow 106$
$\neg 25$	$\rightarrow 107$
$\neg 26$	$\rightarrow 109, \rightarrow 110, \rightarrow 111, \rightarrow 112, \rightarrow 113$
$\neg 27$	$\rightarrow 114, \rightarrow 115, \rightarrow 116, \rightarrow 117, \rightarrow 118$
$\neg 28$	$\rightarrow 119, \rightarrow 120, \rightarrow 121, \rightarrow 122, \rightarrow 123$
$\neg 29$	$\rightarrow 126$
$\neg 30$	$\rightarrow 128$
$\neg 31$	$\rightarrow 131$
$\neg 32$	$\rightarrow 133$
$\neg 33$	$\rightarrow 136$
$\neg 34$	$\rightarrow 138$
$\neg 35$	$\rightarrow 139$
$\neg 36$	$\rightarrow 140$

(continued)

Table 1.3 (continued)

$\neg 37$	$\rightarrow 141$
$\neg 38$	$\rightarrow 142, \rightarrow 143$
$\neg 39$	$\rightarrow 144, \rightarrow 145$
$\neg 40$	$\rightarrow 146, \rightarrow 147$
$\neg 41$	$\rightarrow 148, \rightarrow 149$
$\neg 42$	$\rightarrow 150$
$\neg 43$	$\rightarrow 151$
$\neg 44$	$\rightarrow 152$
$\neg 45$	$\rightarrow 153$
$\neg 46, \lambda$	$\rightarrow 154, \lambda, \rightarrow 155, \lambda$
$\neg 47, \lambda$	$\rightarrow 156, \lambda, \rightarrow 157, \lambda$
$\neg 48, \gamma$	$\rightarrow 158, \gamma, \rightarrow 159, \gamma$
$\neg 49, \gamma$	$\rightarrow 160, \gamma, \rightarrow 161, \gamma$
$\neg 50, \alpha, \beta$	$\rightarrow 162, \alpha, \beta, \rightarrow 163, \alpha, \beta$
$\neg 51, \alpha, \beta$	$\rightarrow 164, \alpha, \beta, \rightarrow 165, \alpha, \beta$
$\neg 52$	$\rightarrow 167$
$\neg 53$	$\rightarrow 176$

$$= (\langle a, b \rangle \vee \neg_8 \langle a, b \rangle) \rightarrow_{34} (\langle a, b \rangle \rightarrow_{34} \neg_8 \neg_8 \langle a, b \rangle)$$

$$= (\langle a, b \rangle \vee \langle 1 - a, a \rangle) \rightarrow_{34} (\langle a, b \rangle \rightarrow_{34} \neg_8 \langle 1 - a, a \rangle)$$

$$= \langle \max(a, 1 - a), \min(a, b) \rangle \rightarrow_{34} (\langle a, b \rangle \rightarrow_{34} \langle a, 1 - a \rangle)$$

$$= \langle \max(a, 1 - a), \min(a, b) \rangle \rightarrow_{34} \langle \min(1, 2 - a - 1 + a), \max(0, a + 1 - a - 1) \rangle$$

$$= \langle \max(a, 1 - a), \min(a, b) \rangle \rightarrow_{34} \langle 1, 0 \rangle$$

$$= \langle \min(1, 2 - a), \max(0, a + 0 - 1) \rangle = \langle 1, 0 \rangle,$$

and (using above calculated results)

$$V((\neg\neg A \vee \neg_8 A) \rightarrow_{34} (A \rightarrow_{34} \neg_8 \neg_8 A))$$

$$= (\neg_8 \neg_8 \langle a, b \rangle \vee \neg_8 \langle a, b \rangle) \rightarrow_{34} (\langle a, b \rangle \rightarrow_{34} \neg_8 \neg_8 \langle a, b \rangle)$$

$$= (\langle a, 1 - a \rangle \vee \langle 1 - a, a \rangle) \rightarrow_{34} \langle 1, 0 \rangle$$

$$= \langle \max(a, 1 - a), \min(a, 1 - a) \rangle \rightarrow_{34} \langle 1, 0 \rangle$$

$$= \langle \min(1, 2 - \max(a, 1 - a)), \max(0, \max(a, 1 - a) + 0 - 1) \rangle = \langle 1, 0 \rangle,$$

i.e., the two formulas have equal values for implication $\to_{34}$ and for negation $\neg_8$ generated by it. Therefore, they satisfy (1.5.7).

On the other hand,

$$V((A \vee \neg_{10} A) \to_{59} (A \to_{59} \neg_{10} \neg_{10} A))$$

$$= (\langle a, b \rangle \vee \neg_{10} \langle a, b \rangle) \to_{59} (\langle a, b \rangle \to_{59} \neg_{10} \neg_{10} \langle a, b \rangle)$$

$$= (\langle a, b \rangle \vee \langle \overline{sg}(1 - b), 1 - b \rangle) \to_{59} (\langle a, b \rangle \to_{59} \neg_{10} \langle \overline{sg}(1 - b), 1 - b \rangle)$$

$$= \langle \max(a, \overline{sg}(1 - b)), \min(b, 1 - b) \rangle \to_{59} (\langle a, b \rangle \to_{59} \langle \overline{sg}(b), b \rangle)$$

$$= \langle \max(a, \overline{sg}(1 - b)), \min(b, 1 - b) \rangle \to_{59} \langle \max(\overline{sg}(1 - b), \overline{sg}(b)), 1 - \max(b, b) \rangle$$

$$= \langle \max(\overline{sg}(1 - \min(b, 1 - b)), \max(\overline{sg}(1 - b), \overline{sg}(b))),$$

$$1 - \max(\min(b, 1 - b), \max(\overline{sg}(1 - b), \overline{sg}(b))) \rangle.$$

If $b = 1$ or $b = 0$, then

$$V((A \vee \neg_{10} A) \to_{59} (A \to_{59} \neg_{10} \neg_{10} A)) = \langle 1, 0 \rangle.$$

If $0 < b < 1$, then

$$V((A \vee \neg_{10} A) \to_{59} (A \to_{59} \neg_{10} \neg_{10} A))$$

$$= \langle \max(\overline{sg}(1 - \min(b, 1 - b)), 0), 1 - \max(\min(b, 1 - b), 0) \rangle = \langle 0, 1 \rangle.$$

Also,

$$V(\neg_{10} \neg_{10} A \vee \neg_{10} A) \to_{59} (A \to_{59} \neg_{10} \neg_{10} A))$$

$$= (\neg_{10} \langle \overline{sg}(1 - b), 1 - b \rangle \vee \langle \overline{sg}(1 - b), 1 - b \rangle) \to_{59} (\langle a, b \rangle \to_{59} \neg_{10} \langle \overline{sg}(1 - b), 1 - b \rangle)$$

$$= (\langle \overline{sg}(b), b \rangle \vee \langle \overline{sg}(1 - b), 1 - b \rangle) \to_{59} (\langle a, b \rangle \to_{59} \langle \overline{sg}(b), b \rangle)$$

$$= \langle \max(\overline{sg}(b), \overline{sg}(1 - b)), \min(b, 1 - b) \rangle \to_{59} \langle \max(\overline{sg}(1 - b), \overline{sg}(b)), 1 - b \rangle$$

$$= \langle \max(\overline{sg}(1 - \min(b, 1 - b)), \max(\overline{sg}(1 - b), \overline{sg}(b))),$$

$$1 - \max(\min(b, 1 - b), \max(\overline{sg}(1 - b), \overline{sg}(b))) \rangle$$

If $b = 1$ or $b = 0$, then

$$V((\neg_{10}(\neg_{10}A \vee \neg_{10}A) \rightarrow_{59} (A \rightarrow_{59} \neg_{10}\neg_{10}A)) = \langle 1, 0 \rangle.$$

If $0 < b < 1$, then

$$V(\neg_{10}\neg_{10}A \vee \neg_{10}A) \rightarrow_{59} (A \rightarrow_{59} \neg_{10}\neg_{10}A))$$

$$= \langle \max(0, 0), 1 - \max(\min(b, 1 - b), 0) \rangle$$

$$= \langle 0, 1 - \min(b, 1 - b) \rangle.$$

Therefore, both formulas have different values and equality (1.5.7) is not valid. The same holds for implications $\rightarrow_{58}, \ldots, \rightarrow_{60}, \rightarrow_{70}, \rightarrow_{72}, \rightarrow_{73}, \rightarrow_{78}, \rightarrow_{80}, \rightarrow_{87}, \rightarrow_{89}, \rightarrow_{92}, \rightarrow_{96}, \rightarrow_{98}, \rightarrow_{120}, \rightarrow_{140}, \rightarrow_{142}, \ldots, \rightarrow_{145}, \rightarrow_{157}, \rightarrow_{162}, \rightarrow_{163}, \rightarrow_{165}.$

On the other hand, formulas (1.5.3) and (1.5.4) have equal behaviour, as illustrated by the following two assertions.

Theorem 1.5.5 *Formula A satisfies (1.5.3) and (1.5.4) as a tautology for the intuitionistic fuzzy implications* $\rightarrow_2, \rightarrow_3, \rightarrow_8, \rightarrow_{11}, \rightarrow_{14}, \ldots, \rightarrow_{16}, \rightarrow_{19}, \rightarrow_{20}, \rightarrow_{23}, \rightarrow_{24}, \rightarrow_{31}, \rightarrow_{32}, \rightarrow_{37}, \rightarrow_{40}, \ldots, \rightarrow_{45}, \rightarrow_{47}, \rightarrow_{48}, \rightarrow_{52}, \rightarrow_{55}, \ldots, \rightarrow_{57}, \rightarrow_{62}, \rightarrow_{65}, \rightarrow_{74}, \rightarrow_{77}, \rightarrow_{83}, \rightarrow_{88}, \rightarrow_{90}, \rightarrow_{97}, \rightarrow_{99}, \rightarrow_{153}, \rightarrow_{171}, \ldots, \rightarrow_{180}.$

Theorem 1.5.6 *Formula A satisfies (1.5.3) and (1.5.4) as an IFT for the intuitionistic fuzzy implications* $\rightarrow_1, \ldots, \rightarrow_9, \rightarrow_{11}, \ldots, \rightarrow_{38}, \rightarrow_{40}, \ldots, \rightarrow_{53}, \rightarrow_{55}, \ldots, \rightarrow_{57}, \rightarrow_{61}, \ldots, \rightarrow_{66}, \rightarrow_{68}, \rightarrow_{69}, \rightarrow_{71}, \rightarrow_{72}, \rightarrow_{74}, \ldots, \rightarrow_{77}, \rightarrow_{79}, \ldots, \rightarrow_{85}, \rightarrow_{88}, \ldots, \rightarrow_{91}, \rightarrow_{93}, \rightarrow_{94}, \rightarrow_{97}, \rightarrow_{100}, \ldots, \rightarrow_{107}, \rightarrow_{109}, \ldots, \rightarrow_{121}, \rightarrow_{122}, \rightarrow_{137}, \rightarrow_{151}, \rightarrow_{153}, \rightarrow_{158}, \ldots, \rightarrow_{161}, \rightarrow_{166}, \ldots, \rightarrow_{185}.$

In all these cases, both formulas satisfy or do not satisfy each of these two formulas. Analogously, we can check that equality

$$V((A \rightarrow \neg\neg A) \rightarrow (A \vee \neg A)) = V((A \rightarrow \neg\neg A) \rightarrow (\neg\neg A \vee \neg A)) \quad (1.5.8)$$

is not valid for implications $\rightarrow_{70}, \rightarrow_{72}, \rightarrow_{73}, \rightarrow_{78}, \rightarrow_{80}, \rightarrow_{84}, \rightarrow_{98}, \rightarrow_{155}, \rightarrow_{156}, \rightarrow_{163}.$

Now, we check the validity of the following two assertions.

Theorem 1.5.7 *Formula A satisfies (1.5.5) as a tautology for the intuitionistic fuzzy implications* $\rightarrow_{20}, \rightarrow_{23}, \rightarrow_{42}, \rightarrow_{74}, \rightarrow_{77}, \rightarrow_{88}, \rightarrow_{90}, \rightarrow_{153}.$

Theorem 1.5.8 *Formula A satisfies (1.5.5) as an IFT for the intuitionistic fuzzy implications* $\rightarrow_1, \rightarrow_4, \ldots, \rightarrow_7, \rightarrow_9, \rightarrow_{13}, \rightarrow_{17}, \rightarrow_{19}, \rightarrow_{20}, \ldots, \rightarrow_{23}, \rightarrow_{25}, \rightarrow_{27}, \ldots, \rightarrow_{30}, \rightarrow_{33}, \ldots, \rightarrow_{36}, \rightarrow_{38}, \rightarrow_{42}, \rightarrow_{44}, \rightarrow_{45}, \rightarrow_{61}, \rightarrow_{64}, \rightarrow_{66}, \rightarrow_{67}, \rightarrow_{71}, \rightarrow_{72}, \rightarrow_{74}, \ldots, \rightarrow_{77}, \rightarrow_{279}, \ldots, \rightarrow_{82}, \rightarrow_{85}, \rightarrow_{86}, \rightarrow_{88}, \ldots, \rightarrow_{90}, \rightarrow_{100}, \ldots, \rightarrow_{105}, \rightarrow_{117}, \rightarrow_{118}, \rightarrow_{124}, \ldots, \rightarrow_{133}, \rightarrow_{151}, \rightarrow_{153}, \rightarrow_{158}, \ldots, \rightarrow_{161}, \rightarrow_{166}, \ldots, \rightarrow_{170}, \rightarrow_{181}, \ldots, \rightarrow_{183}, \rightarrow_{185}.$

In comparison, formulas (1.5.5) and (1.5.6) behave differently. Now, Theorems 1.5.7 and 1.5.8 are changed to the following forms.

Theorem 1.5.9 *Formula A satisfies (1.5.6) as a tautology for the intuitionistic fuzzy implications* $\rightarrow_2, \rightarrow_3, \rightarrow_8, \rightarrow_{11}, \rightarrow_{14}, \ldots, \rightarrow_{16}, \rightarrow_{19}, \rightarrow_{20}, \rightarrow_{23}, \rightarrow_{31}, \rightarrow_{32}, \rightarrow_{37}, \rightarrow_{40}, \ldots, \rightarrow_{45}, \rightarrow_{47}, \rightarrow_{48}, \rightarrow_{52}, \rightarrow_{55}, \ldots, \rightarrow_{57}, \rightarrow_{62}, \rightarrow_{65}, \rightarrow_{74}, \rightarrow_{77}, \rightarrow_{83}, \rightarrow_{88}, \rightarrow_{90}, \rightarrow_{97}, \rightarrow_{153}, \rightarrow_{171}, \ldots, \rightarrow_{180}.$

Theorem 1.5.10 *Formula A satisfies (1.5.6) as an IFT for the intuitionistic fuzzy implications* $\rightarrow_1, \ldots, \rightarrow_9, \rightarrow_{11}, \ldots, \rightarrow_{38}, \rightarrow_{40}, \ldots, \rightarrow_{53}, \rightarrow_{55}, \ldots, \rightarrow_{57}, \rightarrow_{61}, \rightarrow_{62}, \rightarrow_{64}, \ldots, \rightarrow_{67}, \rightarrow_{71}, \rightarrow_{72}, \rightarrow_{74}, \ldots, \rightarrow_{77}, \rightarrow_{79}, \ldots, \rightarrow_{83}, \rightarrow_{85}, \rightarrow_{86}, \rightarrow_{88}, \ldots, \rightarrow_{91}, \rightarrow_{94}, \rightarrow_{95}, \rightarrow_{97}, \rightarrow_{98}, \rightarrow_{100}, \ldots, \rightarrow_{107}, \rightarrow_{109}, \ldots, \rightarrow_{137}, \rightarrow_{151}, \rightarrow_{153}, \rightarrow_{158}, \ldots, \rightarrow_{161}, \rightarrow_{166}, \ldots, \rightarrow_{185}.$

Having in mind the definitions of tautology and IFT (see (1.1.7) and (1.1.8)), let us define for every evaluation function V and for formulas A and B:

$$V(A) = V(B) \text{ if and only if } \mu(A) = \mu(B) \text{ and } \nu(A) = \nu(B).$$

Let $A \equiv_\rightarrow B$ denote $(A \rightarrow B) \wedge (B \rightarrow A)$ for any fixed implication $\rightarrow$. For example, when $\rightarrow$ is $\rightarrow_4$, we prove the following lemma.

Lemma 1.5.1 *If $V(A) = V(B)$, then, $A \equiv_\rightarrow B$ is an IFT.*

Proof From $V(A) = V(B)$ and from:

$$V(A \equiv_\rightarrow B) = \langle \min(\max(\nu(A), \mu(B)), \max(\mu(A), \nu(B))),$$

$$\max(\min(\nu(A), \mu(B)), \min(\mu(A), \nu(B))) \rangle$$

it follows, that:

$$\begin{aligned}
&\min(\max(\nu(A), \mu(B)), \max(\mu(A), \nu(B))) \\
&- \max(\min(\nu(A), \mu(B)), \min(\mu(A), \nu(B))) \\
=\ &\min(\max(\nu(A), \mu(A)), \max(\mu(A), \nu(A))) \\
&- \max(\min(\nu(A), \mu(A)), \min(\mu(A), \nu(A))) \\
=\ &\max(\nu(A), \mu(A)) - \min(\nu(A), \mu(A)) \geq 0,
\end{aligned}$$

i.e., $A \equiv_\rightarrow B$ is an IFT.

The opposite assertion is not valid. For example, if $V(A) = \langle 0.4, 0.5 \rangle$ and $V(B) = \langle 0.4, 0.3 \rangle$ then $V(A \equiv B) = \langle 0.4, 0.4 \rangle$, i.e., $A \equiv_\rightarrow B$ is an IFT, but obviously, $V(A) \neq V(B)$.   ☐

Open Problem 1. Determine for which indices i and any two formulas A and B it is valid that:

$A \equiv_\rightarrow B$ is a tautology, if and only if $V(A) = V(B)$?

The weak form of this Problem is whether for the same conditions,

$A \equiv_\rightarrow B$ is an IFT if and only if $V(A) = V(B)$.

Open Problem 2. Determine for which implications formulas (1.5.2)–(1.5.6) are tautologies (or IFTs).

In [47], inspired by [66], the following formulas are studied:

$$(A \wedge B) \rightarrow C \equiv_\rightarrow (A \rightarrow (B \rightarrow C)), \tag{1.5.9}$$
$$A \rightarrow B \equiv_\rightarrow (A \rightarrow (A \rightarrow B)), \tag{1.5.10}$$

and for them, the following two theorems are proved.

Theorem 1.5.11 *Implications* $\rightarrow_3, \rightarrow_4, \rightarrow_{11}, \rightarrow_{12}, \rightarrow_{14}, \rightarrow_{16}, \rightarrow_{18}, \ldots, \rightarrow_{20}, \rightarrow_{22}, \rightarrow_{23}, \rightarrow_{25}, \ldots, \rightarrow_{28}, \rightarrow_{31}, \ldots, \rightarrow_{33}, \rightarrow_{41}, \ldots, \rightarrow_{43}, \rightarrow_{48}, \rightarrow_{56}, \rightarrow_{57}, \rightarrow_{74}, \rightarrow_{76}, \rightarrow_{77}, \rightarrow_{79}, \rightarrow_{81}, \rightarrow_{88}, \rightarrow_{97} \overset{*}{} \rightarrow_{153}, \rightarrow_{171}, \ldots, \rightarrow_{175}$ *satisfy (1.5.9) as tautologies.*

Theorem 1.5.12 *Implications* $\rightarrow_1, \ldots, \rightarrow_4, \rightarrow_8, \rightarrow_{10}, \ldots, \rightarrow_{12}, \rightarrow_{14}, \rightarrow_{16}, \ldots, \rightarrow_{20}, \rightarrow_{22}, \rightarrow_{23}, \rightarrow_{25}, \ldots, \rightarrow_{28}, \rightarrow_{30}, \ldots, \rightarrow_{33}, \rightarrow_{36}, \rightarrow_{37}, \rightarrow_{39}, \ldots, \rightarrow_{43}, \rightarrow_{48}, \rightarrow_{51}, \rightarrow_{52}, \rightarrow_{54}, \ldots, \rightarrow_{57}, \rightarrow_{59}, \rightarrow_{61}, \rightarrow_{67}, \rightarrow_{72}, \ldots, \rightarrow_{74}, \rightarrow_{76}, \ldots \rightarrow_{81}, \rightarrow_{86}, \ldots, \rightarrow_{89}, \rightarrow_{91}, \rightarrow_{92}, \rightarrow_{95}, \ldots, \rightarrow_{97}, \rightarrow_{100}, \rightarrow_{105}, \rightarrow_{106}, \rightarrow_{109}, \rightarrow_{110}, \rightarrow_{114}, \rightarrow_{119}, \rightarrow_{120}, \overset{*}{} \rightarrow_{153}, \rightarrow_{166}, \rightarrow_{168}, \rightarrow_{171}, \ldots, \rightarrow_{180}$ *satisfy (1.5.10) as tautologies.*

In Theorems 1.5.11–1.5.14, the lists of implications before symbol "*" are published by L. Atanassova and after this symbol they are added by the author.

In [48], the following formulas are also studied:

$$(A \vee B) \rightarrow C = (A \rightarrow C) \wedge (B \rightarrow C), \tag{1.5.11}$$
$$(A \wedge B) \rightarrow C = (A \rightarrow C) \vee (B \rightarrow C). \tag{1.5.12}$$

Theorem 1.5.13 *Implications* $\rightarrow_2, \rightarrow_3, \rightarrow_4, \rightarrow_5, \rightarrow_{12}, \rightarrow_{13}, \rightarrow_{14}, \rightarrow_{16}, \rightarrow_{18}, \rightarrow_{19}, \rightarrow_{20}, \rightarrow_{22}, \rightarrow_{23}, \rightarrow_{25}, \rightarrow_{26}, \rightarrow_{27}, \rightarrow_{28}, \rightarrow_{29}, \rightarrow_{31}, \rightarrow_{32}, \rightarrow_{33}, \rightarrow_{34}, \rightarrow_{35}, \rightarrow_{37}, \rightarrow_{40}, \rightarrow_{41}, \rightarrow_{42}, \rightarrow_{43}, \rightarrow_{44}, \rightarrow_{45}, \rightarrow_{47}, \rightarrow_{48}, \rightarrow_{49}, \rightarrow_{50}, \rightarrow_{52}, \rightarrow_{55}, \rightarrow_{56}, \rightarrow_{57}, \rightarrow_{58}, \rightarrow_{59}, \rightarrow_{60}, \rightarrow_{61}, \rightarrow_{64}, \rightarrow_{66}, \rightarrow_{67}, \rightarrow_{69}, \rightarrow_{70}, \rightarrow_{71}, \rightarrow_{72}, \rightarrow_{73}, \rightarrow_{74}, \rightarrow_{76}, \rightarrow_{77}, \rightarrow_{78}, \rightarrow_{79}, \rightarrow_{80}, \rightarrow_{81}, \rightarrow_{82}, \rightarrow_{83}, \rightarrow_{84}, \rightarrow_{85}, \rightarrow_{86}, \rightarrow_{87}, \rightarrow_{88}, \rightarrow_{89}, \rightarrow_{90}, \rightarrow_{91}, \rightarrow_{92}, \rightarrow_{93}, \rightarrow_{94}, \rightarrow_{95}, \rightarrow_{96}, \rightarrow_{97}, \rightarrow_{98}, \rightarrow_{99}, \rightarrow_{102}, \rightarrow_{105}, \rightarrow_{108}, \rightarrow_{124}, \rightarrow_{125}, \rightarrow_{127}, \rightarrow_{129}, \rightarrow_{130}, \rightarrow_{132}, \rightarrow_{134}, \rightarrow_{135}, \rightarrow_{137}$ *satisfy (1.5.11) as tautologies.*

Theorem 1.5.14 *Implications* $\rightarrow_2, \rightarrow_3, \rightarrow_4, \rightarrow_5, \rightarrow_8, \rightarrow_{11}, \rightarrow_{12}, \rightarrow_{13}, \rightarrow_{16}, \rightarrow_{18}, \rightarrow_{19}, \rightarrow_{20}, \rightarrow_{22}, \rightarrow_{23}, \rightarrow_{25}, \rightarrow_{26}, \rightarrow_{27}, \rightarrow_{28}, \rightarrow_{29}, \rightarrow_{31}, \rightarrow_{32}, \rightarrow_{33}, \rightarrow_{34}, \rightarrow_{35}, \rightarrow_{37}, \rightarrow_{40}, \rightarrow_{41}, \rightarrow_{42}, \rightarrow_{43}, \rightarrow_{44}, \rightarrow_{45}, \rightarrow_{47}, \rightarrow_{48}, \rightarrow_{49}, \rightarrow_{50}, \rightarrow_{52}, \rightarrow_{55}, \rightarrow_{56}, \rightarrow_{57}, \rightarrow_{58}, \rightarrow_{59}, \rightarrow_{60}, \rightarrow_{61}, \rightarrow_{62}, \rightarrow_{63}, \rightarrow_{64}, \rightarrow_{65}, \rightarrow_{66}, \rightarrow_{67}, \rightarrow_{68}, \rightarrow_{70}, \rightarrow_{71},$

$\rightarrow_{72}$, $\rightarrow_{73}$, $\rightarrow_{74}$, $\rightarrow_{76}$, $\rightarrow_{77}$, $\rightarrow_{78}$, $\rightarrow_{79}$, $\rightarrow_{80}$, $\rightarrow_{81}$, $\rightarrow_{82}$, $\rightarrow_{83}$, $\rightarrow_{84}$, $\rightarrow_{85}$, $\rightarrow_{86}$, $\rightarrow_{87}$, $\rightarrow_{88}$, $\rightarrow_{89}$, $\rightarrow_{90}$, $\rightarrow_{91}$, $\rightarrow_{92}$, $\rightarrow_{93}$, $\rightarrow_{94}$, $\rightarrow_{95}$, $\rightarrow_{96}$, $\rightarrow_{97}$, $\rightarrow_{98}$, $\rightarrow_{99}$, $\rightarrow_{102}$, $\rightarrow_{105}$, $\rightarrow_{108}$, $\rightarrow_{124}$, $\rightarrow_{125}$, $\rightarrow_{127}$, $\rightarrow_{129}$, $\rightarrow_{130}$, $\rightarrow_{132}$, $\rightarrow_{134}$, $\rightarrow_{135}$, $\rightarrow_{137}$ *satisfy (1.5.12) as tautologies.*

Now, we check which intuitionistic fuzzy implications and negations satisfy C.A. Meredith's axiom (see, e.g., [2]).

Theorem 1.5.15 *For every five formulas A, B, C, D and E, Meredith's axiom*

$$((((A \rightarrow B) \rightarrow (\neg C \rightarrow \neg D)) \rightarrow C) \rightarrow E) \rightarrow ((E \rightarrow A) \rightarrow (D \rightarrow A))$$

is valid as a tautology by implications $\rightarrow_{20}$, $\rightarrow_{23}$, $\rightarrow_{74}$, $\rightarrow_{77}$, $\rightarrow_{153}$.

Theorem 1.5.16 *For every five formulas A, B, C, D and E, Meredith's axiom is valid as an IFT by implications* $\rightarrow_1$, $\rightarrow_4$, $\rightarrow_5$, $\rightarrow_6$, $\rightarrow_9$, $\rightarrow_{13}$, $\rightarrow_{17}$, $\rightarrow_{18}$, $\rightarrow_{20}$, $\ldots$, $\rightarrow_{23}$, $\rightarrow_{25}$, $\rightarrow_{27}$, $\ldots$, $\rightarrow_{29}$, $\rightarrow_{61}$, $\rightarrow_{64}$, $\rightarrow_{72}$, $\rightarrow_{74}$, $\ldots$, $\rightarrow_{77}$, $\rightarrow_{79}$, $\ldots$, $\rightarrow_{81}$, $\rightarrow_{100}$, $\ldots$, $\rightarrow_{102}$, $\rightarrow_{109}$, $\ldots$, $\rightarrow_{113}$, $\rightarrow_{124}$, $\ldots$, $\rightarrow_{128}$, $\rightarrow_{133}$, $\rightarrow_{151}$, $\rightarrow_{153}$, $\rightarrow_{158}$, $\ldots$, $\rightarrow_{161}$, $\rightarrow_{166}$, $\rightarrow_{167}$, $\rightarrow_{169}$, $\rightarrow_{170}$, $\rightarrow_{182}$, $\rightarrow_{185}$

Proof Let $V(A) = \langle a, b \rangle$, $V(B) = \langle c, d \rangle$, $V(C) = \langle e, f \rangle$, $V(D) = \langle g, h \rangle$, $V(E) = \langle i, j \rangle$, where $a, b, \ldots, j \in [0, 1]$ and $a + b \leq 1, c + d \leq 1, e + f \leq 1, g + h \leq 1$ and $i + j \leq 1$. We check the validity of the assertion for the case when the implication is, for example, $\rightarrow_4$.

$$V((((A \rightarrow_4 B) \rightarrow_4 (\neg C \rightarrow_4 \neg D)) \rightarrow_4 C) \rightarrow_4 E) \rightarrow_4 ((E \rightarrow_4 A) \rightarrow_4 (D \rightarrow_4 A)))$$

$$= (((((\langle a, b \rangle \rightarrow_4 \langle c, d \rangle) \rightarrow_4 (\langle f, e \rangle \rightarrow_4 \langle h, g \rangle)) \rightarrow_4 \langle e, f \rangle) \rightarrow_4 \langle i, j \rangle)$$
$$\rightarrow_4 (((\langle i, j \rangle \rightarrow_4 \langle a, b \rangle) \rightarrow_4 (\langle g, h \rangle \rightarrow_4 \langle a, b \rangle)))$$
$$= (((((\langle \max(b, c), \min(a, d) \rangle \rightarrow_4 \langle \max(e, h), \min(f, g) \rangle)$$
$$\rightarrow_4 \langle e, f \rangle) \rightarrow_4 \langle i, j \rangle) \rightarrow_4 (((\langle i, j \rangle \rightarrow_4 \langle a, b \rangle) \rightarrow_4 (\langle g, h \rangle \rightarrow_4 \langle a, b \rangle)))$$
$$= (((\langle \max(e, h, \min(a, d)), \min(f, g, \max(b, c)) \rangle \rightarrow_4 \langle e, f \rangle)$$
$$\rightarrow_4 \langle i, j \rangle) \rightarrow_4 (((\langle i, j \rangle \rightarrow_4 \langle a, b \rangle) \rightarrow_4 (\langle g, h \rangle \rightarrow_4 \langle a, b \rangle)))$$
$$= ((\langle \max(e, \min(f, g, \max(b, c))), \min(f, \max(e, h, \min(a, d))) \rangle$$
$$\rightarrow_4 \langle i, j \rangle) \rightarrow_4 ((\langle \max(a, j), \min(b, i) \rangle \rightarrow_4 \langle \max(a, h), \min(b, g) \rangle))$$
$$= \langle \max(i, \min(f, \max(e, h, \min(a, d)))), \min(j, \max(e, \min(f, g,$$
$$\max(b, c)))) \rangle \rightarrow_4 \langle \max(a, h, \min(b, i)), \min(b, g, \max(a, j)) \rangle$$
$$= \langle \max(a, h, \min(b, i), \min(j, \max(e, \min(f, g, \max(b, c))))),$$
$$\min(b, g, \max(a, j), \max(i, \min(f, \max(e, h, \min(a, d))))) \rangle.$$

Let

$$X = \max(a, h, \min(b, i), \min(j, \max(e, \min(f, g, \max(b, c))))) -$$
$$\min(b, g, \max(a, j), \max(i, \min(f, \max(e, h, \min(a, d))))).$$

Obviously,

$$\max(a, b) \ge \min(a, b), \max(a, b) \ge \min(b, j), \max(a, i) \ge \min(a, b).$$

Let $\max(a, i) \ge \min(b, j)$. Then,

$$\begin{aligned}
X &\ge \max(a, \min(b, i)) - \min(b, \max(a, j)) \\
&= \min(\max(a, b), \max(a, i)) - \max(\min(a, b), \min(b, j)) \ge 0.
\end{aligned}$$

Let $\max(a, i) < \min(b, j)$. Then, $a < b, a < j, i < b$ and $i < j$.
If $j \le \max(e, \min(f, g, \max(b, c)))$, then

$$\begin{aligned}
X &= \max(a, h, i, \min(j, \max(e, \min(f, g, \max(b, c)))))- \\
&\qquad \min(b, g, j, \max(i, \min(f, \max(e, h, \min(a, d))))) \\
&\ge \max(a, h, i, j) - \min(b, g, j) \ge 0.
\end{aligned}$$

If $i \ge \min(f, \max(e, h, \min(a, d)))$, then

$$\begin{aligned}
X &= \max(a, h, i, \min(j, \max(e, \min(f, g, \max(b, c)))))- \\
&\qquad \min(b, g, j, \max(i, \min(f, \max(e, h, \min(a, d))))) \\
&\ge \max(a, h, i) - \min(b, g, i, j) \ge 0.
\end{aligned}$$

Finally, let

$$j > \max(e, \min(f, g, \max(b, c)))$$

and

$$i < \min(f, \max(e, h, \min(a, d))).$$

Then, $j > e$ and

$$j > \min(f, g, \max(b, c))$$

and $i < f$ and

$$i < \max(e, h, \min(a, d)).$$

Therefore,

$$\begin{aligned}
X &= \max(a, h, i.e., \min(f, g, \max(b, c))) \\
&\quad - \min(b, g, j, f, \max(e, h, \min(a, d))) \\
&\ge \min(f, g, \max(b, c))) - \min(b, g, j, f, \max(e, h, \min(a, d))) \\
&\ge \min(b, f, g) - \min(b, g, j, f, \max(e, h, \min(a, d))) \ge 0.
\end{aligned}$$

Therefore, the Meredith's axiom is an IFT.   □

The validity of the axioms of the intuitionistic logic (IL) (see, e.g., [67]) is checked
for all implications in [28].

Below, we give the list of the implications for which all axioms of IL are valid.
The checks are done both for the case of tautologies, as well as for the case of IFTs.

The IL axioms are the following.

(IL1) $A \rightarrow A$,

(IL2) $A \rightarrow (B \rightarrow A)$,

(IL3) $A \rightarrow (B \rightarrow (A \wedge B))$,

(IL4) $(A \rightarrow (B \rightarrow C)) \rightarrow (B \rightarrow (A \rightarrow C))$,

(IL5) $(A \rightarrow (B \rightarrow C)) \rightarrow ((A \rightarrow B) \rightarrow (A \rightarrow C))$,

(IL6) $A \rightarrow \neg\neg A$,

(IL7) $\neg(A \wedge \neg A)$,

(IL8) $(\neg A \vee B) \rightarrow (A \rightarrow B)$,

(IL9) $\neg(A \vee B) \rightarrow (\neg A \wedge \neg B)$,

(IL10) $(\neg A \wedge \neg B) \rightarrow \neg(A \vee B)$,

(IL11) $(\neg A \vee \neg B) \rightarrow \neg(A \wedge B)$,

(IL12) $(A \rightarrow B) \rightarrow (\neg B \rightarrow \neg A)$,

(IL13) $(A \rightarrow \neg B) \rightarrow (B \rightarrow \neg A)$,

(IL14) $\neg\neg\neg A \rightarrow \neg A$,

(IL15) $\neg A \rightarrow \neg\neg\neg A$,

(IL16) $\neg\neg(A \rightarrow B) \rightarrow (A \rightarrow \neg\neg B)$,

(IL17) $(C \rightarrow A) \rightarrow ((C \rightarrow (A \rightarrow B)) \rightarrow (C \rightarrow B))$.

Theorem 1.5.17 *Implications* $\rightarrow_1, \rightarrow_3, \ldots, \rightarrow_5, \rightarrow_9, \rightarrow_{11}, \rightarrow_{13}, \rightarrow_{14}, \rightarrow_{17}, \rightarrow_{18}, \rightarrow_{20}, \rightarrow_{21}, \rightarrow_{22}, \rightarrow_{23}, \rightarrow_{27}, \ldots, \rightarrow_{29}, \rightarrow_{61}, \rightarrow_{66}, \rightarrow_{71}, \rightarrow_{74}, \ldots, \rightarrow_{77}, \rightarrow_{79}, \rightarrow_{81}, \rightarrow_{100}, \ldots, \rightarrow_{102}, \rightarrow_{109}, \ldots, \rightarrow_{113}, \rightarrow_{118}, \rightarrow_{124}, \ldots, \rightarrow_{128}, \rightarrow_{151}, \rightarrow_{153}, \rightarrow_{158}, \ldots, \rightarrow_{160}, \rightarrow_{166}, \rightarrow_{167}, \rightarrow_{169}, \rightarrow_{170}, \rightarrow_{182}, \rightarrow_{185}$ *satisfy all IL axioms as IFTs.*

Proof Let us assume everywhere below that

$$V(A) = \langle a, b \rangle$$

$$V(B) = \langle c, d \rangle$$

$$V(C) = \langle e, f \rangle$$

Then, the validity of (IL5) for implication $\rightarrow_4$ is checked as follows. $V((A \rightarrow (B \rightarrow C)) \rightarrow ((A \rightarrow B) \rightarrow (A \rightarrow C)))$

$= (\langle a, b \rangle \rightarrow \langle \max(d, e), \min(c, f) \rangle) \rightarrow (\langle \max(b, c), \min(a, d) \rangle$
$\quad \rightarrow \langle \max(b, e), \min(a, f) \rangle)$

$= \langle \max(b, d, e), \min(a, c, f) \rangle \rightarrow \langle \max(b, e, \min(a, d)), \min(a, f, \max(b, c)) \rangle$

$= \langle \max(b, e, \min(a, d), \min(a, c, f)), \min(a, f, \max(b, c), \max(b, d, e)) \rangle$

and

$\max(b, e, \min(a, d), \min(a, c, f)) \geq \max(b, e, \min(a, d))$
$\geq \min(a, \max(b, d, e)) \geq \min(a, f, \max(b, c), \max(b, d, e))$.

The validity of the other axioms is checked analogically. $\qquad \square$

Theorem 1.5.18 *Only implication $\rightarrow_{153}$ satisfies all IL axioms as tautologies.*

The validity of Kolmogorov's and Łukasiewicz–Tarski's Axioms of Logic (see, e.g., [68]) are checked for all implications in [22] .

Now, we give the lists of the implications for which all Kolmogorov's and Łukasiewicz–Tarski's axioms are valid. The checks are done as for the case of tautologies, as well as for the case of IFTs.

The first group of axioms (of Kolmogorov) comprises

(K1) $A \rightarrow (B \rightarrow A)$,

(K2) $(A \rightarrow (A \rightarrow B)) \rightarrow (A \rightarrow B))$,

(K3) $(A \rightarrow (B \rightarrow C)) \rightarrow (B \rightarrow (A \rightarrow C))$,

(K4) $(B \rightarrow C) \rightarrow ((A \rightarrow B) \rightarrow (A \rightarrow C))$,

(K5) $(A \rightarrow B) \rightarrow ((A \rightarrow \neg B) \rightarrow \neg A)$.

Theorem 1.5.19 *Implications $\rightarrow_1, \rightarrow_3, \ldots, \rightarrow_5, \rightarrow_9, \rightarrow_{11}, \rightarrow_{13}, \rightarrow_{14}, \rightarrow_{17}, \rightarrow_{18}, \rightarrow_{20}, \rightarrow_{22}, \rightarrow_{23}, \rightarrow_{27}, \ldots, \rightarrow_{29}, \rightarrow_{61}, \rightarrow_{66}, \rightarrow_{71}, \rightarrow_{74}, \ldots, \rightarrow_{77}, \rightarrow_{79}, \rightarrow_{81}, \rightarrow_{100}, \ldots, \rightarrow_{102}, \rightarrow_{109}, \ldots, \rightarrow_{113}, \rightarrow_{118}, \rightarrow_{124}, \ldots, \rightarrow_{128}, \rightarrow_{151}, \rightarrow_{153}, \rightarrow_{158}, \ldots, \rightarrow_{160}, \rightarrow_{166}, \rightarrow_{167}, \rightarrow_{169}, \rightarrow_{170}, \rightarrow_{182}, \rightarrow_{185}$ satisfy all Kolmogorov's axioms as IFTs.*

Theorem 1.5.20 *Implications $\rightarrow_3, \rightarrow_{11}, \rightarrow_{14}, \rightarrow_{20}, \rightarrow_{23}, \rightarrow_{74}, \rightarrow_{77}, \rightarrow_{153}$ satisfy all Kolmogorov's axioms as tautologies.*

The second group of axioms (of Łukasiewicz and Tarski) is

(LT1) $A \rightarrow (B \rightarrow A)$,

(LT2) $(A \rightarrow B) \rightarrow ((B \rightarrow C) \rightarrow (A \rightarrow C))$,

(LT3) $\neg A \rightarrow (\neg B \rightarrow (B \rightarrow A))$,

(LT4) $((A \rightarrow \neg A) \rightarrow A) \rightarrow A$.

Theorem 1.5.21 *Implications* $\rightarrow_1, \rightarrow_4, \ldots, \rightarrow_6, \rightarrow_9, \rightarrow_{17}, \rightarrow_{18}, \rightarrow_{20}, \ldots, \rightarrow_{23},$ $\rightarrow_{27}, \ldots, \rightarrow_{29}, \rightarrow_{61}, \rightarrow_{64}, \rightarrow_{66}, \rightarrow_{71}, \rightarrow_{74}, \ldots, \rightarrow_{77}, \rightarrow_{79}, \rightarrow_{81}, \rightarrow_{100}, \ldots, \rightarrow_{102},$ $\rightarrow_{109}, \ldots, \rightarrow_{113}, \rightarrow_{124}, \ldots, \rightarrow_{128}, \rightarrow_{153}$ *satisfy all Łukasiewicz–Tarski's axioms as IFTs.*

Theorem 1.5.22 *Implications* $\rightarrow_{20}, \rightarrow_{23}, \rightarrow_{74}, \rightarrow_{77}, \rightarrow_{153}$ *satisfy all Łukasiewicz–Tarski's axioms as tautologies.*

Sixth, some variants of fuzzy implications (marked by $I(x, y)$) are described in the book of Klir and Yuan [56] and the following nine axioms are discussed, where $I(x, y)$ denotes $x \rightarrow y$ for any of the possible forms of the operation implication, N is the operation negation related with operation $\rightarrow$, and for $a, b, c, d \in [0, 1], a+b \leq 1, c + d \leq 1$:

$$\langle a, b \rangle \leq \langle c, d \rangle \text{ iff } a \leq c \text{ and } b \geq d. \qquad (1.5.13)$$

Axiom $A1$ $(\forall x, y)(x \leq y \rightarrow (\forall z)(I(x, z) \geq I(y, z)))$,
Axiom $A2$ $(\forall x, y)(x \leq y \rightarrow (\forall z)(I(z, x) \leq I(z, y)))$,
Axiom $A3$ $(\forall y)(I(0, y) = 1)$,
Axiom $A4$ $(\forall y)(I(1, y) = y)$,
Axiom $A5$ $(\forall x)(I(x, x) = 1)$,
Axiom $A6$ $(\forall x, y, z)(I(x, I(y, z)) = I(y, I(x, z)))$,
Axiom $A7$ $(\forall x, y)(I(x, y) = 1 \text{ iff } x \leq y)$,
Axiom $A8$ $(\forall x, y)(I(x, y) = I(N(y), N(x)))$,
Axiom $A9$ I is a continuous function.

For our research, having in mind the specific forms of the intuitionistic fuzzy implications, we modify five of these axioms, as follows.
Axiom $A3^*$ $(\forall y)(I(0, y)$ is an IFT),
Axiom $A4^*$ $(\forall y)(I(1, y) \leq y)$,
Axiom $A5^*$ $(\forall x)(I(x, x)$ is an IFT),
Axiom $A7^*$ $(\forall x, y)$ (if $x \leq y$, then, $I(x, y) = 1$),
Axiom $A8^*$ $(\forall x, y)(I(x, y) = N(N(I(N(y), N(x)))))$.

Here, we ignore Axiom 9, because, obviously, it is valid for all the implications that do not contain operations sg or $\overline{sg}$.

Following the paper of N. Angelova and the author [21], we formulate the following theorem.

Theorem 1.5.23 *The intuitionistic fuzzy implications that satisfy Klir and Yuan's axioms as (standard) tautologies, are marked in Table 1.4 by "$\bullet$" and the implications that satisfy the same axioms (only) as IFTs – by "$\circ$".*

Finally, following [8], we discuss the well-known Contraposition Law

$$(A \rightarrow B) \rightarrow (\neg B \rightarrow \neg A) \qquad (1.5.14)$$

and its modified version

$$(\neg\neg A \rightarrow \neg\neg B) \rightarrow (\neg B \rightarrow \neg A). \tag{1.5.15}$$

For them, the following assertions are valid.

Theorem 1.5.24 *(a) Implications* $\rightarrow_1, \ldots, \rightarrow_9, \rightarrow_{11}, \ldots, \rightarrow_{29}, \rightarrow_{46}, \ldots, \rightarrow_{53}, \rightarrow_{55},$
$\ldots, \rightarrow_{57}, \rightarrow_{61}, \rightarrow_{62}, \rightarrow_{64}, \ldots, \rightarrow_{66}, \rightarrow_{71}, \rightarrow_{72}, \rightarrow_{74}, \ldots, \rightarrow_{77}, \rightarrow_{79}, \ldots, \rightarrow_{81},$
$\rightarrow_{91}, \rightarrow_{94}, \rightarrow_{99}, \rightarrow_{100}, \ldots, \rightarrow_{102}, \rightarrow_{104}, \rightarrow_{105}, \rightarrow_{109}, \ldots, \rightarrow_{113}, \rightarrow_{118}, \rightarrow_{120},$
$\ldots, \rightarrow_{122}, \rightarrow_{124}, \ldots, \rightarrow_{128}, \rightarrow_{133}, \ldots, \rightarrow_{137}, \rightarrow_{151}, \rightarrow_{153}, \rightarrow_{158}, \ldots, \rightarrow_{161}, \rightarrow_{166},$
$\rightarrow_{167}, \rightarrow_{169}, \qquad \ldots, \rightarrow_{172}, \rightarrow_{174}, \ldots, \rightarrow_{177}, \rightarrow_{180}, \ldots, \rightarrow_{182}, \rightarrow_{184}, \rightarrow_{185}$
satisfy (1.5.14) as IFTs.
(b) Implications $\rightarrow_2, \rightarrow_3, \rightarrow_8, \rightarrow_{11}, \rightarrow_{14}, \ldots, \rightarrow_{16}, \rightarrow_{19}, \rightarrow_{20}, \rightarrow_{23}, \rightarrow_{24}, \rightarrow_{47},$
$\ldots, \rightarrow_{49}, \rightarrow_{52}, \rightarrow_{55}, \ldots, \rightarrow_{57}, \rightarrow_{65}, \rightarrow_{74}, \rightarrow_{77}, \rightarrow_{97}, \rightarrow_{153}, \rightarrow_{171}, \rightarrow_{172}, \rightarrow_{174},$
$\ldots, \rightarrow_{177}, \rightarrow_{179}, \rightarrow_{180}, \rightarrow_{182}, \rightarrow_{184}, \rightarrow_{185}$ *satisfy (1.5.14) as tautologies.*

Theorem 1.5.25 *(a) Implications* $\rightarrow_1, \ldots, \rightarrow_9, \rightarrow_{11}, \ldots, \rightarrow_{38}, \rightarrow_{40}, \ldots, \rightarrow_{53}, \rightarrow_{55},$
$\ldots, \rightarrow_{57}, \rightarrow_{61}, \rightarrow_{62}, \rightarrow_{64}, \ldots, \rightarrow_{66}, \rightarrow_{71}, \rightarrow_{72}, \rightarrow_{74}, \ldots, \rightarrow_{77}, \rightarrow_{79}, \ldots, \rightarrow_{83},$
$\rightarrow_{85}, \rightarrow_{88}, \ldots, \rightarrow_{91}, \rightarrow_{94}, \rightarrow_{97}, \rightarrow_{100}, \ldots, \rightarrow_{107}, \rightarrow_{109}, \ldots, \rightarrow_{118}, \rightarrow_{120}, \ldots,$
$\rightarrow_{122}, \rightarrow_{124}, \ldots, \rightarrow_{137}, \rightarrow_{151}, \rightarrow_{153}, \rightarrow_{158}, \ldots, \rightarrow_{161}, \rightarrow_{166}, \ldots, \rightarrow_{185}$ *satisfy*
(1.5.15) as IFTs.
(b) Implications $\rightarrow_2, \rightarrow_3, \rightarrow_8, \rightarrow_{11}, \rightarrow_{14}, \ldots, \rightarrow_{16}, \rightarrow_{19}, \rightarrow_{20}, \rightarrow_{23}, \rightarrow_{24}, \rightarrow_{31},$
$\rightarrow_{32}, \rightarrow_{34}, \rightarrow_{34}, \rightarrow_{41}, \ldots, \rightarrow_{45}, \rightarrow_{47}, \ldots, \rightarrow_{49}, \rightarrow_{52}, \rightarrow_{55}, \ldots, \rightarrow_{57}, \rightarrow_{65}, \rightarrow_{74},$
$\rightarrow_{77}, \rightarrow_{83}, \rightarrow_{88}, \rightarrow_{90}, \rightarrow_{97}, \rightarrow_{153}, \rightarrow_{171}, \ldots, \rightarrow_{180}, \rightarrow_{182}, \ldots, \rightarrow_{185}$ *satisfy*
(1.5.15) as tautologies.

Theorem 1.5.26 *(a) Implications* $\rightarrow_1, \rightarrow_4, \rightarrow_5, \rightarrow_7, \rightarrow_9, \rightarrow_{13}, \rightarrow_{18}, \rightarrow_{20}, \ldots,$
$\rightarrow_{23}, \rightarrow_{25}, \rightarrow_{27}, \ldots, \rightarrow_{29}, \rightarrow_{61}, \rightarrow_{66}, \rightarrow_{71}, \rightarrow_{74}, \rightarrow_{76}, \rightarrow_{77}, \rightarrow_{79}, \rightarrow_{81}, \rightarrow_{100},$
$\ldots, \rightarrow_{102}, \rightarrow_{104}, \rightarrow_{105}, \rightarrow_{107}, \rightarrow_{109}, \ldots, \rightarrow_{113}, \rightarrow_{118}, \rightarrow_{124}, \ldots, \rightarrow_{128}, \rightarrow_{133},$
$\rightarrow_{151}, \rightarrow_{153}, \rightarrow_{158}, \ldots, \rightarrow_{161}, \rightarrow_{166}, \rightarrow_{167}, \rightarrow_{169}, \rightarrow_{170}, \rightarrow_{182}, \rightarrow_{185}$ *satisfy*
expression

$$(\neg A \rightarrow \neg B) \rightarrow ((\neg A \rightarrow B) \rightarrow A)$$

as IFTs.
(b) Implications $\rightarrow_{20}, \rightarrow_{23}, \rightarrow_{74}, \rightarrow_{77}, \rightarrow_{153}$ *satisfy this expression as tautologies.*

Theorem 1.5.27 *(a) Implications* $\rightarrow_1, \rightarrow_4, \rightarrow_5, \rightarrow_7, \rightarrow_9, \rightarrow_{13}, \rightarrow_{18}, \rightarrow_{20}, \ldots,$
$\rightarrow_{23}, \rightarrow_{25}, \rightarrow_{27}, \ldots, \rightarrow_{30}, \rightarrow_{33}, \ldots, \rightarrow_{36}, \rightarrow_{38}, \rightarrow_{42}, \rightarrow_{45}, \rightarrow_{61}, \rightarrow_{66}, \rightarrow_{71},$
$\rightarrow_{74}, \rightarrow_{76}, \rightarrow_{77}, \rightarrow_{79}, \rightarrow_{81}, \rightarrow_{82}, \rightarrow_{85}, \rightarrow_{88}, \rightarrow_{100}, \ldots, \rightarrow_{105}, \rightarrow_{107}, \rightarrow_{109},$
$\ldots, \rightarrow_{118}, \rightarrow_{124}, \ldots, \rightarrow_{133}, \rightarrow_{151}, \rightarrow_{153}, \rightarrow_{158}, \ldots, \rightarrow_{161}, \rightarrow_{166}, \ldots, \rightarrow_{170},$
$\rightarrow_{182}, \rightarrow_{183}, \rightarrow_{185}$ *satisfy expression*

$$(\neg A \rightarrow \neg B) \rightarrow ((\neg A \rightarrow \neg\neg B) \rightarrow A)$$

as IFTs.
(b) Implications $\rightarrow_{20}, \rightarrow_{23}, \rightarrow_{42}, \rightarrow_{45}, \rightarrow_{74}, \rightarrow_{77}, \rightarrow_{88}, \rightarrow_{153}$ *satisfy the same*
expression as tautologies.

Table 1.4 The intuitionistic fuzzy implications that satisfy Klir and Yuan's axioms

	A1	A2	A3	A3*	A4	A4*	A5	A5*	A6	A7	A7*	A8	A8*
1		•	•	•	•	○		○					
2	•	•	•	•		•	•	•			•		
3	•	•	•	•	•	•	•	•	•		•		
4	•	•	•	•	•	•		○	•			•	•
5	•	•	•	•	•	•		○	•			•	•
6		•	•	•	•	•		○					
7				○	•	•		○				•	○
8	•	•	•	•		•	•	•			•		
9		•	•	•	•	•		○					
10		•	•	•	•	•							
11	•	•	•	•	•	•	•	•	•		•		
12	•	•	•	•		•		•					
13	•	•	•	•	•	•		○	•			•	•
14	•	•	•	•	•	•	•	•	•	•	•		
15	•	•	•	•		•	•	•		•	•		
16	•	•	•	•	•	•		•					
17		•	•	•	•	•		○	•				
18	•	•	•	•	•	•		○	•				
19	•	•	•	•	•	•			•				
20	•	•	•	•			•	•	•		•	•	•
21			•	•				○					
22	•	•	•	•				○	•			•	•
23	•	•	•	•			•	•	•		•	•	•
24	•	•	•	•		•	•	•		•	•		
25	•	•	•	•		•			•				
26	•	•	•	•	•	•			•				
27	•	•	•	○				○	•			•	•
28	•	•	•	•	•	•		○	•				
29	•	•	•	•				○					
30		•	•	•				○					
31	○	○	○	○			•	•	•		•		
32	•	•	•	•			•	•	•		•		
33	•	•	•	•				○	•				
34	•	•	•	•			•	•	•		•		
35	•	•	•	•				○	•				
36				○				○	•				
37	•	•	•	•			•	•			•		
38		•	•	•				○					
39		•	•	•									
40	•	•	•	•			•	•			•		

(continued)

Table 1.4 (continued)

	A1	A2	A3	A3*	A4	A4*	A5	A5*	A6	A7	A7*	A8	A8*
41	•	•	•	•					•				
42	•	•	•	•			•	•	•		•		
43	•	•	•	•					•				
44	•	•	•	•				○					
45	•	•	•	•				○					
46		•	•	•		•							
47	•	•	•	•		•							
48	•	•	•	•		•			•				
49	•	•	•	•		•			•				
50	•	•	•	•		•			•				
51				○		•			•				
52	•	•	•	•		•							
53		•	•	•		•							
54		•	•	•		•							
55	•	•	•	•		•							
56	•	•	•	•		○			•				
57	•	•	•	•					•				
58	•	•	•	•		•							
59	•	•	•	•		•							
60	•	•	•	•									
61	•			○	•	○		○					
62	•	•	•	•			•	•			•		
63	•	•	•	•			•	•			•		
64	•			○	•	•		○					
65	•	•	•	•			•	•			•		
66	•			○				○					
67	•		•	•	•	•							
68	•	•	•	•			•	•			•		
69	•	•	•	•		•	•	•		•	•		
70	•	•	•	•									
71	•		•	•				○					
72	•	•	•	•		•		○					
73	•	•	•	•		•							
74	•	•	•	•			•	•	•		•	•	•
75			•	•				○					
76	•	•	•	•		•		○	•			•	•
77	•	•	•	•		•	•	•	•		•	•	•
78	•	•	•	•		•							
79	•	•	•	•		•		○	•			•	•

(continued)

Table 1.4 (continued)

	A1	A2	A3	A3*	A4	A4*	A5	A5*	A6	A7	A7*	A8	A8*
80	•	•	•	•		•		○					
81	•	•	•	•		•		○	•				
82	•			○				○					
83	•	•	•	•			•	•			•		
84	•	•	•	•			•	•			•		
85	•			○				○					
86	•		•	•									
87	•	•	•	•									
88	•	•	•	•			•	•	•		•		
89	•	•	•	•				○					
90	•	•	•	•				○					
91	•			○	•								
92	•	•	•	•	•								
93	•	•	•	•	•								
94	•			○	•								
95	•		•	•	•								
96	•	•	•	•	•								
97	•	•	•	•	•			•					
98	•	•	•	•	•								
99	•	•	•	•	•								
100	○	○		○				○	•		○		○
101	○	○		○				○			○		○
102	○	○	•	○				○			○		○
103	○	○		○				○			○		○
104	○	○		○				○			○		○
105	○	○	•	○				○			○		○
106	○	○		○							○		○
107	○	○		○							○		○
108	○	○	•	•							○		○
109		•	•	•	•	•		○					○
110		•	•	•	•	•		○	•				○
111			•	•				○					○
112		•	•	•	•	•		○	•				○
113				○				○					○
114		•	•	•				○					○
115		•	•	○				○					○
116			•	•				○					○
117		•	•	•				○					○
118				○				○					○

(continued)

Table 1.4 (continued)

	A1	A2	A3	A3*	A4	A4*	A5	A5*	A6	A7	A7*	A8	A8*
119		•	•	•		•							○
120		•	•	•		•							○
121			•	•									○
122		•	•	•		•							○
123				○									○
124	•			○				○					○
125	•		•	•				○					○
126			•	•				○					○
127	•		•	•				○					○
128				○				○					○
129	•			○				○					○
130	•		•	•				○					○
131			•	•				○					○
132	•		•	•				○					○
133				○				○					○
134	•			○									○
135	•		•	•									○
136			•	•									○
137	•		•	•									○
138				○									○
139	○	○		○		•		○					○
140						•							○
141	○	○		○		•		○					○
142													○
143						•							○
144													○
145						•							○
146	○	○		○				○					○
147	○	○		○		•		○					○
148	○	○		○				○					○
149	○	○		○		•							○
150	○	○		○				○					○
151	○	○		○				○					○
152	○	○		○				○					○
153	•	•	•	•			•	•	•		•	•	•
154	○	○		○				○					○
155	○	○		○				○					○
156	○	○		○									○
157	○	○		○				○					○
158	○	○		○				○					○

(continued)

Table 1.4 (continued)

	A1	A2	A3	A3*	A4	A4*	A5	A5*	A6	A7	A7*	A8	A8*
159	○	○		○				○					○
160	○	○		○				○					○
161	○	○		○									○
162	○	○		○				○					○
163	○	○		○				○					○
164	○	○		○				○					○
165	○	○		○									○
166		●	●	●	●	●		○					○
167	○	●	●	●				○			○		○
168		●	●	●				○					○
169	○	●	●	○				○			○		○
170		●	●	●				○					○
171	●	●	●	●	●	●			●				○
172	●	●	●	●		●			●				○
173	●	●	●	●					●				○
174	●	●	●	●					●				○
175	●	●	●	●		●			●				○
176	●	●	●	●		●	●	●			●		○
177	●	●	●	●		●	●	●			○		●
178	●	●	●	●		●	●	●			●		○
179	●	●	●	●		●							○
180	●	●	●	●			●	●			●		○
181	●	●	●	●	●	●	●	●			●		○
182	●	●	●	●		●	●	●	●		●	●	●
183	●	●	●	●			●	●	●		●		○
184	●	●	●	●		●			●				○
185	●	●	●	●	○	○	●	●	●	○	●	●	●

Theorem 1.5.28 (a) *Implications* $\to_1, \ldots, \to_5, \to_7, \ldots, \to_9, \to_{11}, \ldots, \to_{29},$ $\to_{46}, \ldots, \to_{53}, \to_{55}, \ldots, \to_{57}, \to_{61}, \to_{66}, \to_{71}, \to_{74}, \ldots, \to_{77}, \to_{79}, \to_{81}, \to_{91},$ $\to_{94}, \to_{97}, \to_{99}, \ldots, \to_{102}, \to_{104}, \ldots, \to_{107}, \to_{109}, \ldots, \to_{113}, \to_{118}, \to_{119},$ $\to_{121}, \to_{124}, \ldots, \to_{128}, \to_{133}, \ldots, \to_{137}, \to_{151}, \to_{153}, \to_{158}, \ldots, \to_{161}, \to_{166},$ $\to_{167}, \to_{169}, \ldots, \to_{172}, \to_{174}, \ldots, \to_{177}, \to_{179}, \to_{180}, \to_{182}, \to_{184}, \to_{185}$ *satisfy expression*

$$(\neg A \to \neg B) \to ((\neg A \to B) \to \neg\neg A)$$

as IFTs.
(b) *Implications* $\to_2, \to_3, \to_8, \to_{11}, \to_{14}, \to_{15}, \to_{16}, \to_{19}, \to_{20}, \to_{23}, \to_{47},$

$\rightarrow_{48}, \rightarrow_{52}, \rightarrow_{55}, \ldots, \rightarrow_{57}, \rightarrow_{74}, \rightarrow_{77}, \rightarrow_{97}, \rightarrow_{99}, \rightarrow_{153}, \rightarrow_{171}, \rightarrow_{172}, \rightarrow_{174},$
$\ldots, \rightarrow_{177}, \rightarrow_{179}, \rightarrow_{180}$ *satisfy the same expression as tautologies.*

Theorem 1.5.29 *(a)   Implications* $\rightarrow_1, \ldots, \rightarrow_5, \rightarrow_7, \ldots, \rightarrow_9, \rightarrow_{11}, \ldots, \rightarrow_{38},$
$\rightarrow_{40}, \ldots, \rightarrow_{43}, \rightarrow_{45}, \ldots, \rightarrow_{53}, \rightarrow_{55}, \ldots, \rightarrow_{57}, \rightarrow_{61}, \rightarrow_{66}, \rightarrow_{71}, \rightarrow_{74}, \ldots, \rightarrow_{77},$
$\rightarrow_{79}, \rightarrow_{81}, \ldots, \rightarrow_{83}, \rightarrow_{85}, \rightarrow_{88}, \rightarrow_{91}, \rightarrow_{94}, \rightarrow_{97}, \rightarrow_{99}, \ldots, \rightarrow_{119}, \rightarrow_{121}, \rightarrow_{124},$
$\ldots, \rightarrow_{137}, \rightarrow_{151}, \rightarrow_{153}, \rightarrow_{158}, \ldots, \rightarrow_{161}, \rightarrow_{166}, \ldots, \rightarrow_{180}, \rightarrow_{182}, \ldots, \rightarrow_{185}$ *satisfy expression*

$$(\neg A \rightarrow \neg B) \rightarrow ((\neg A \rightarrow \neg\neg B) \rightarrow \neg\neg A).$$

as IFTs.
(b)   Implications $\rightarrow_2, \rightarrow_3, \rightarrow_8, \rightarrow_{11}, \rightarrow_{14}, \ldots, \rightarrow_{16}, \rightarrow_{19}, \rightarrow_{20}, \rightarrow_{23}, \rightarrow_{31},$
$\rightarrow_{32}, \rightarrow_{37}, \rightarrow_{40}, \ldots, \rightarrow_{43}, \rightarrow_{45}, \rightarrow_{47}, \rightarrow_{48}, \rightarrow_{52}, \rightarrow_{55}, \ldots, \rightarrow_{57}, \rightarrow_{74}, \rightarrow_{77},$
$\rightarrow_{83}, \rightarrow_{88}, \rightarrow_{97}, \rightarrow_{99}, \rightarrow_{153}, \rightarrow_{171}, \ldots, \rightarrow_{180}$ *satisfy the same expression as tautologies.*

In [69, 70], the Hauber's Law is formulated by

$$((A \rightarrow B) \wedge (C \rightarrow D) \wedge (A \vee C) \wedge \neg(B \wedge D)) \rightarrow ((B \rightarrow A) \wedge (D \rightarrow C))$$

and it is proved that it is a standard tautology.

Here we shall prove the following theorem.

Theorem 1.5.30 *The Hauber's Law is an IFT for standard intuitionistic fuzzy implication* $(\rightarrow_4)$, *negation* $(\neg_1)$, *disjunction and conjunction.*

Proof Let the formulas A, B, C, D be given and let everywhere:

$$V(A) = \langle a, b \rangle,$$

$$V(B) = \langle c, d \rangle,$$

$$V(C) = \langle e, f \rangle,$$

$$V(D) = \langle g, h \rangle.$$

Then, following [71] we calculate sequentially

$$V(((A \rightarrow B) \wedge (C \rightarrow D) \wedge (A \vee C) \wedge \neg(B \wedge D)) \rightarrow ((B \rightarrow A) \wedge (D \rightarrow C)))$$

$$= (((\langle a, b \rangle \rightarrow \langle c, d \rangle) \wedge (\langle e, f \rangle \rightarrow \langle g, h \rangle) \wedge (\langle a, b \rangle \vee \langle e, f \rangle)$$

$$\wedge \neg(\langle c, d \rangle \wedge \langle g, h \rangle)) \rightarrow ((\langle c, d \rangle \rightarrow \langle a, b \rangle) \wedge (\langle g, h \rangle \rightarrow \langle e, f \rangle)))$$

$$= (\langle \max(b, c), \min(a, d) \rangle \wedge \langle \max(f, g), \min(e, h) \rangle \wedge \langle \max(a, e), \min(b, f) \rangle$$

$$\wedge \langle \max(d, h), \min(c, g) \rangle) \rightarrow (\langle \max(a, d), \min(b, c) \rangle \wedge \langle \max(e, h), \min(f, g) \rangle)$$

$$= (\langle \min(\max(b, c), \max(f, g), \max(a, e), \max(d, h)), \max(\min(a, d), \min(e, h),$$

$$\min(b, f), \min(c, g)))) \rightarrow \langle \min(\max(a, d), \max(e, h)), \max(\min(b, c), \min(f, g)) \rangle$$

$$= \langle \max(\min(a, d), \min(e, h), \min(b, f), \min(c, g), \min(\max(a, d), \max(e, h))),$$

$$\min(\max(b, c), \max(f, g), \max(a, e), \max(d, h), \max(\min(b, c), \min(f, g)))) \rangle.$$

Let

$$X \equiv \max(\min(a, d), \min(e, h), \min(b, f), \min(c, g), \min(\max(a, d), \max(e, h)))$$

$$- \min(\max(b, c), \max(f, g), \max(a, e), \max(d, h), \max(\min(b, c), \min(f, g)))$$

$$\geq \max(\min(a, d), \min(e, h), \min(\max(a, d), \max(e, h)))$$

$$- \min(\max(a, e), \max(d, h)).$$

If $a \geq d \geq e \geq h$, then $X \geq \max(d, h, \min(a, e)) - \min(a, d) = d - d = 0$.

If $a \geq d \geq h \geq e$, then $X \geq \max(d, e, \min(a, h)) - \min(a, d) = d - d = 0$.

If $a \geq e \geq d \geq h$, then $X \geq \max(d, h, \min(a, e)) - \min(a, d) = d - d = 0$.

If $a \geq e \geq h \geq d$, then $X \geq \max(d, h, \min(a, e)) - \min(a, h) = e - h \geq 0$.

If $a \geq h \geq d \geq e$, then $X \geq \max(d, e, \min(a, h)) - \min(a, h) = h - h = 0$.

If $a \geq h \geq e \geq d$, then $X \geq \max(d, e, \min(a, h)) - \min(a, h) = h - h = 0$.

If $d \geq a \geq e \geq h$, then $X \geq \max(a, h, \min(d, e)) - \min(a, d) = a - a = 0$.

If $d \geq a \geq h \geq e$, then $X \geq \max(a, e, \min(d, h)) - \min(a, d) = a - a = 0$.

If $d \geq e \geq a \geq h$, then $X \geq \max(a, h, \min(d, e)) - \min(e, d) = e - e = 0$.

If $d \geq e \geq h \geq a$, then $X \geq \max(a, h, \min(d, e)) - \min(e, d) = e - e = 0$.

If $d \geq h \geq a \geq e$, then $X \geq \max(a, e, \min(d, h)) - \min(a, d) = h - a \geq 0$.

If $d \geq h \geq e \geq a$, then $X \geq \max(a, e, \min(d, h)) - \min(e, d) = h - e \geq 0$.

If $e \geq a \geq d \geq h$, then $X \geq \max(d, h, \min(a, e)) - \min(e, d) = a - d \geq 0$.

If $e \geq a \geq h \geq d$, then $X \geq \max(d, h, \min(a, e)) - \min(e, h) = a - h \geq 0$.

If $e \geq d \geq a \geq h$, then $X \geq \max(a, h, \min(d, e)) - \min(e, d) = d - d = 0$.

If $e \geq d \geq h \geq a$, then $X \geq \max(a, h, \min(d, e)) - \min(e, d) = d - d = 0$.

If $e \geq h \geq a \geq d$, then $X \geq \max(d, h, \min(a, e)) - \min(e, h) = h - h = 0$.

If $e \geq h \geq d \geq a$, then $X \geq \max(a, h, \min(d, e)) - \min(e, h) = h - h = 0$.

If $h \geq a \geq e \geq d$, then $X \geq \max(d, e, \min(a, h)) - \min(a, h) = a - a = 0$.

If $h \geq a \geq d \geq e$, then $X \geq \max(d, e, \min(a, h)) - \min(a, h) = a - a = 0$.

If $h \geq d \geq a \geq e$, then $X \geq \max(a, e, \min(d, h)) - \min(a, h) = d - a \geq 0$.

If $h \geq d \geq e \geq a$, then $X \geq \max(a, e, \min(d, h)) - \min(e, h) = d - e \geq 0$.

If $h \geq e \geq a \geq d$, then $X \geq \max(d, e, \min(a, h)) - \min(e, h) = e - e = 0$.

If $h \geq e \geq d \geq a$, then $X \geq \max(a, e, \min(d, h)) - \min(e, h) = e - e = 0$.
Therefore, the Hauber's Law is an IFT. □

This theorem generates the following open problem.

Open Problem 3. For which other intuitionistic fuzzy implications, negations, disjunctions and conjunctions the Hauber's Law is an IFT? Are there operations of these four types, for which the Law is a standard tautology?

The expression

$$((A \vee B) \wedge (\neg A \vee C)) \rightarrow (B \vee C) \tag{1.5.16}$$

is called "the basic axiom of the resolution" (see [72]). Obviously, it is a tautology in the sense of first-order logic. Here we discuss its intuitionistic fuzzy interpretation.

Theorem 1.5.31 *For every three formulas A, B and C,*
(a) (1.5.16) is an IFT for implications $\rightarrow_1, \ldots, \rightarrow_5, \rightarrow_8, \rightarrow_{11}, \rightarrow_{13}, \ldots, \rightarrow_{15}, \rightarrow_{17}, \rightarrow_{18}, \rightarrow_{20}, \ldots, \rightarrow_{24}, \rightarrow_{27}, \ldots, \rightarrow_{38}, \rightarrow_{40}, \rightarrow_{42}, \rightarrow_{44}, \rightarrow_{45}, \rightarrow_{61}, \rightarrow_{65}, \rightarrow_{68}, \rightarrow_{69}, \rightarrow_{71}, \rightarrow_{74}, \ldots, \rightarrow_{77}, \rightarrow_{79}, \rightarrow_{81}, \ldots, \rightarrow_{85}, \rightarrow_{88}, \ldots, \rightarrow_{90}, \rightarrow_{100}, \ldots, \rightarrow_{105}, \rightarrow_{109}, \ldots, \rightarrow_{118}, \rightarrow_{124}, \ldots, \rightarrow_{133}, \rightarrow_{151}, \rightarrow_{158}, \ldots, \rightarrow_{160}, \rightarrow_{166}, \ldots, \rightarrow_{170}, \rightarrow_{176}, \ldots, \rightarrow_{178}, \rightarrow_{180}, \ldots, \rightarrow_{183}, \rightarrow_{185}.$
(b) (1.5.16) is a tautology for implications $\rightarrow_2, \rightarrow_8, \rightarrow_{11}, \rightarrow_{14}, \rightarrow_{15}, \rightarrow_{20}, \rightarrow_{23}, \rightarrow_{31}, \rightarrow_{32}, \ldots, \rightarrow_{37}, \rightarrow_{40}, \rightarrow_{42}, \rightarrow_{74}, \rightarrow_{77}, \rightarrow_{83}, \rightarrow_{88}, \rightarrow_{176}, \rightarrow_{177}, \rightarrow_{178}, \rightarrow_{180}.$

Proof Following [73], we check the validity of (a) for implication $\rightarrow_4$. Let $V(A) = \langle a, b \rangle$, $V(B) = \langle c, d \rangle$, $V(C) = \langle e, f \rangle$, where $a, b, c, d, e, f \in [0, 1], a + b \leq 1, c + d \leq 1, e + f \leq 1$. Then,

$$V(((A \vee B)) \wedge ((\neg A \vee C)) \rightarrow_4 (B \vee C))$$

$$= (\langle \max(a, c), \min(b, d) \rangle \wedge \langle \max(b, e), \min(a, f) \rangle) \rightarrow_4 \langle \max(c, e), \min(d, f) \rangle$$

$$= \langle \min(\max(a, c), \max(b, e)), \max(\min(b, d), \min(a, f)) \rangle$$

$$\rightarrow_4 \langle \max(c, e), \min(d, f) \rangle$$

$$= \langle \max(c, e, \min(b, d), \min(a, f)), \min(d, f, \max(a, c), \max(b, e)) \rangle.$$

Then,

$$\max(c, e, \min(b, d), \min(a, f)) - \min(d, f, \max(a, c), \max(b, e))$$

$$\geq \max(c, \min(a, f)) - \min(f, \max(a, c)) \geq 0,$$

i.e., (1.5.16) is an IFT. $\square$

Here, we give a modification of the Conditional logic VW (see [74]), discussed in [75]. It uses two different implications, mentioned by $\rightarrow$ and $\supset$. Here, for implication $\rightarrow$ we use implication $\rightarrow_4$ and for implication $\supset$ – implication $\rightarrow_{11}$. In [75], operation "$\rightarrow$" is called "ordinary material conditional" and operation "$\supset$" we use "counterfactual conditional".

Conditional logic contains the following axioms:

Axiom VW1. $p \supset p$

Axiom VW2. $(p \supset (q \rightarrow r)) \rightarrow ((p \supset q) \rightarrow (p \supset r))$

Axiom VW3. $(p \supset q) \rightarrow (p \rightarrow q)$

Axiom VW4. $(\neg p \supset p) \rightarrow (q \supset p)$

Axiom VW5. $((p \supset q) \wedge \neg(p \supset \neg r)) \rightarrow ((p \wedge r) \supset q)$

Axiom VW6. $((p \wedge q) \supset r) \rightarrow (p \supset (q \rightarrow r))$

Axiom VW7. $((p \supset q) \wedge (q \supset p)) \rightarrow ((p \supset r) \equiv (q \supset r))$

and rules:

Rule VW1. From p and $p \rightarrow q$ to infer q

Rule VW2. From $(p_1 \wedge \ldots \wedge p_n) \rightarrow q, r \supset p_1, \ldots, r \supset p_n$ to infer $r \supset q$ (where $n \geq 0$).

Here we shall assume that for the propositions x and y:

$$x \equiv y \text{ iff } (x \supset y) \wedge (y \supset x).$$

Let everywhere below $V(p) = \langle a, b \rangle$, $V(p_i) = \langle a_i, b_i \rangle$ ($1 \leq i \leq n$; n - natural number), $V(q) = \langle c, d \rangle$ and $V(r) = \langle e, f \rangle$.

It is seen directly, that Rule1 is valid, while Rule2 is valid in the following form: if $(p_1 \wedge \ldots \wedge p_n) \rightarrow q$, is true and $r \supset p_1, \ldots, r \supset p_n$ are IF-sure, then, $r \supset q$ is IF-sure (where $n \geq 0$).

Let $(p_1 \wedge \ldots \wedge p_n) \rightarrow q$ is true, and let $r \supset p_1, \ldots, r \supset p_n$ are IF-sure, i.e.,

$$V((p_1 \wedge \ldots \wedge p_n) \rightarrow q) = \langle 1, 0 \rangle,$$

$$\mu(r \supset p_1) \geq \frac{1}{2},$$

$$\ldots$$

$$\mu(r \supset p_n) \geq \frac{1}{2}.$$

Therefore,

$$1 - (1 - c).sg(\min(a_1, \ldots, a_n) - c) = 1,$$

$$d.sg(\min(a_1, \ldots, a_n) - c).sg(d - \max(b_1, \ldots, b_n)) = 0,$$

$$\max(a_1, f) \geq \frac{1}{2},$$

$$\ldots$$

$$\max(a_n, f) \geq \frac{1}{2}.$$

Let us assume that $r \supset q$ is not an IS, i.e.,

$$\max(c, f) < \frac{1}{2}.$$

Therefore, $c < \frac{1}{2}$ and $f < \frac{1}{2}$. Therefore, for every i $(1 \leq i \leq n) : a_i \geq \frac{1}{2}$, i.e.,

$$\min(a_1, \ldots, a_n) \geq \frac{1}{2}.$$

Hence $sg(\min(a_1, \ldots, a_n) - c) = 1$, i.e., $1 - (1 - c) = 1$ and $c = 1$, which is a contradiction.

Therefore, $\max(c, f) \geq \frac{1}{2}$ and hence $r \supset q$ is an IF-sure.

Now, we shall check the validity of the above Axioms.

Axiom VW1 is an IFT, because:

$$V(p \supset p) = \langle 1 - (1 - a).sg(a - a), b.sg(a - a).sg(b - b) \rangle = \langle 1, 0 \rangle.$$

Axiom VW2 is an IFT. This is valid for the following reason.

$V((p \supset (q \rightarrow r)) \rightarrow ((p \supset q) \rightarrow (p \supset r)))$
$= (\langle a, b \rangle \supset \langle 1 - (1 - e).sg(c - e), f.sg(c - e).sg(f - d) \rangle)$
$\quad \rightarrow (\langle \max(b, c), \min(a, d) \rangle \rightarrow \langle \max(b, e), \min(a, f) \rangle)$
$= \langle \max(b, 1 - (1 - e).sg(c - e)), \min(a, f.sg(c - e).sg(f - d)) \rangle$
$\quad \rightarrow \langle 1 - (1 - \max(b, e)).sg(\max(b, c) - \max(b, e)), \min(a, f)$
$\quad .sg(\max(b, c) - \max(b, e)).sg(\min(a, f) - \min(a, d)) \rangle$
$= \langle 1 - (1 - (1 - (1 - \max(b, e)).sg(\max(b, c) - \max(b, e))))$
$\quad .sg(\max(b, 1 - (1 - e).sg(c - e)) - 1 + (1 - \max(b, e))$
$\quad .sg(\max(b, c) - \max(b, e))), \min(a, f).sg(\max(b, c) - \max(b, e))$
$\quad .sg(\min(a, f) - \min(a, d)).sg(\max(b, 1 - (1 - e).sg(c - e)) - 1$
$\quad + (1 - \max(b, e)).sg(\max(b, c) - \max(b, e)))$

$.sg(\min(a, f).sg(\max(b, c) - \max(b, e))$
$.sg(\min(a, f) - \min(a, d)) - \min(a, f.sg(c - e).sg(f - d)))\rangle.$

Let

$A \equiv 1 - (1 - (1 - (1 - \max(b, e)).sg(\max(b, c) - \max(b, e))))$
$.sg(\max(b, 1 - (1 - e).sg(c - e)) - 1 + (1 - \max(b, e))$
$.sg(\max(b, c) - \max(b, e))) - \min(a, f).sg(\max(b, c) - \max(b, e))$
$.sg(\min(a, f) - \min(a, d)).sg(\max(b, 1 - (1 - e).sg(c - e)) - 1$
$+(1 - \max(b, e)).sg(\max(b, c) - \max(b, e))).sg(\min(a, f).sg(\max(b, c)$
$- \max(b, e)).sg(\min(a, f) - \min(a, d)) - \min(a, f.sg(c - e).sg(f - d))).$

Then,

1. If $c \leq e$, or if $c > e$ and $b \geq \max(c, e)$, then, $\max(b, e) \geq \max(b, c)$ and
   $A = 1 - sg(\max(b, 1 - (1 - e).sg(c - e)) - 1) - \min(a, f)$
   $.sg(\max(b, c) - \max(b, e)).sg(\min(a, f) - \min(a, d)).sg(\max(b, 1 - (1 - e)$
   $.sg(c - e)) - 1).sg(\min(a, f).sg(\max(b, c) - \max(b, e))$
   $.sg(\min(a, f) - \min(a, d)) - \min(a, f.sg(c - e).sg(f - d)))$
   $= 1 - 0 = 1;$
2. If $c > \max(b, e)$, then
   $A = sg(\max(b, 1 - (1 - e).sg(c - e)) - 1 + (1 - \max(b, e)).sg(\max(b, c) - \max(b, e)))$
   $= sg(\max(b, 1 - (1 - e)) - 1 + (1 - \max(b, e)))$
   $= sg(\max(b, -e) - \max(b, e)) = 0$
   and
   $A = 0 - 0 = 0.$

Therefore, in both cases $A \geq 0$, i.e.,

$$(p \supset (q \rightarrow r)) \rightarrow ((p \supset q) \rightarrow (p \supset r))$$

is an IFT.

For **Axiom VW3** we obtain

$V((p \supset q) \rightarrow (p \rightarrow q)) = \langle \max(b, c), \min(a, d) \rangle \rightarrow \langle 1 - (1 - c)$
$.sg(a - c), d.sg(a - c).sg(d - b) \rangle$
$= \langle 1 - (1 - (1 - (1 - c).sg(a - c))).sg(\max(b, c) - (1 - (1 - c).sg(a - c))),$
$d.sg(a - c).sg(d - b).sg(\max(b, c) - (1 - (1 - c)$
$.sg(a - c))).sg(d.sg(a - c).sg(d - b) - \min(a, d)) \rangle.$

Let

$A \equiv 1 - (1 - (1 - (1 - c).sg(a - c))).sg(\max(b, c) - (1 - (1 - c).sg(a - c)))$
$-d.sg(a - c).sg(d - b).sg(\max(b, c) - (1 - (1 - c).sg(a - c))).$
$sg(d.sg(a - c).sg(d - b) - \min(a, d)).$

1. If $a \leq c$, then,
   $A = 1 - 0 = 1.$
2. If $a > c$, then
   $A = 1 - (1 - c).sg(\max(b, c) - c) - d.sg(d - b).sg(\max(b, c) - c).sg(d.sg(d - b) - \min(a, d))$

2.1. If $b \leq c$, then,
$$A = 1 - 0 = 1;$$
2.2. If $b > c$, then,
$$A = c - d.\text{sg}(d - b).\text{sg}(d.\text{sg}(d - b) - \min(a, d))$$
  2.2.1. If $d \leq b$, then,
$$A = c - 0 = c;$$
  2.2.2. If $d > b$, then,
$$A = c - d.\text{sg}(d - \min(a, d))$$
    2.2.2.1. If $a \geq d$, then,
$$A = c - 0 = c;$$
    2.2.2.2. If $a < d$, then,
$$A = c - d < 0,$$

because $d > b > c$. Therefore, in this case the Axiom VW3 is not valid.

On the other hand, the following form of Axiom VW3 is valid:

AxiomVW3'. $(p \to q) \supset (p \supset q)$.

In this case:
$$V((p \to q) \supset (p \supset q))$$
$$= \langle 1 - (1 - c).\text{sg}(a - c), d.\text{sg}(a - c).\text{sg}(d - b)\rangle \supset \langle\max(b, c), \min(a, d)\rangle$$
$$= \langle\max(b, c, d.\text{sg}(a - c).\text{sg}(d - b)), \min(a, d, 1 - (1 - c).\text{sg}(a - c))\rangle.$$
Let

$$A \equiv \max(b, c, d.\text{sg}(a - c).\text{sg}(d - b)) - \min(a, d, 1 - (1 - c).\text{sg}(a - c))\rangle.$$

If $a \leq c$, then,
$$A = \max(b, c, 0) - \min(a, d, 1)) \geq c - a \geq 0;$$
If $a > c$, then,
$$A = \max(b, c, d.\text{sg}(d - b)) - \min(a, d, c) \geq c - c = 0,$$
i.e., Axiom VW3' is an IFT.

Axiom VW4 is an IFT, because:
$$V((\neg p \supset p) \to (q \supset p))$$
$$= \langle\max(a, a), \min(b, b)\rangle \to \langle\max(a, d), \min(b, c)\rangle$$
$$= \langle a, b\rangle \to \langle\max(a, d), \min(b, c)\rangle$$
$$= \langle\max(a, b, d), \min(a, b, c)\rangle$$
and obviously,
$$\max(a, b, d) - \min(a, b, c) \geq 0.$$

For **Axiom VW5** we obtain
$$V(((p \supset q) \wedge \neg(p \supset \neg r)) \to ((p \wedge r) \supset q))$$
$$= (\langle\max(b, c), \min(a, d)\rangle \wedge \langle\min(a, e),$$
$$\max(b, f)\rangle) \to ((\langle\min(a, e), \max(b, f)\rangle \to \langle c, d\rangle)$$
$$= \langle\min(a, e, \max(b, c)), \max(b, f, \min(a, d))\rangle \to \langle\max(b, c, f), \min(a, d, e)\rangle$$
$$= \langle 1 - (1 - \max(b, c, f)).\text{sg}(\min(a, e, \max(b, c)) - \max(b, c, f)), \min(a, d, e).$$

$$\text{sg}(\min(a, e, \max(b, c)) - \max(b, c, f)).\text{sg}(\min(a, d, e)$$
$$- \max(b, f, \min(a, d))))\rangle$$
$$= \langle 1, 0 \rangle,$$

because

$$\min(a, e, \max(b, c)) \le \max(b, c,) \le \max(b, c, f))$$

and, therefore,

$$\text{sg}(\min(a, e, \max(b, c)) - \max(b, c, f)) = 0.$$

Therefore,

$$((p \supset q) \wedge \neg(p \supset \neg r)) \rightarrow ((p \wedge r) \supset q))$$

is an IFT.

For **Axiom VW6** we obtain
$$V(((p \wedge q) \supset r) \rightarrow (p \supset (q \rightarrow r)))$$
$$= (\min(a, c), \max(b, d)) \supset \langle e, f \rangle)$$
$$\rightarrow (\langle a, b \rangle \supset \langle 1 - (1 - e).\text{sg}(c - e), f.\text{sg}(c - e).\text{sg}(f - d) \rangle)$$
$$= \langle \max(b, d, e), \min(a, c, f) \rangle \rightarrow \langle \max(b, 1 - (1 - e).\text{sg}(c - e)),$$
$$\min(a, f.\text{sg}(c - e).\text{sg}(f - d)) \rangle$$
$$= \langle 1 - (1 - \max(b, 1 - (1 - e).\text{sg}(c - e)).\text{sg}(\max(b, d, e) - \max(b,$$
$$1 - (1 - e).\text{sg}(c - e))), \min(a, f.\text{sg}(c - e).\text{sg}(f - d)).\text{sg}(\max(b, d, e)$$
$$- \max(b, 1 - (1 - e).\text{sg}(c - e))).\text{sg}(\min(a, f.\text{sg}(c - e)$$
$$.\text{sg}(f - d)) - \min(a, c, f)) \rangle.$$

Let
$$A \equiv 1 - (1 - \max(b, 1 - (1 - e).\text{sg}(c - e))).\text{sg}(\max(b, d, e) - \max(b,$$
$$1 - (1 - e).\text{sg}(c - e))) - \min(a, f.\text{sg}(c - e).\text{sg}(f - d)).\text{sg}(\max(b, d, e)$$
$$- \max(b, 1 - (1 - e).\text{sg}(c - e))).\text{sg}(\min(a, f.\text{sg}(c - e).\text{sg}(f - d))$$
$$- \min(a, c, f)).$$

1. Let $c \le e$, then
   $$A = 1 - (1 - \max(b, 1)).\text{sg}(\max(b, d, e) - \max(b, 1)) - \min(a, 0)$$
   $$.\text{sg}(\max(b, d, e) - \max(b, 1)).\text{sg}(\min(a, 0) - \min(a, c, f))$$
   $$= 1 - 0 = 1;$$
2. Let $c > e$. Then,
   $$A = 1 - (1 - \max(b, e)).\text{sg}(\max(b, d, e) - \max(b, e)) - \min(a, f.\text{sg}(f - d))$$
   $$.\text{sg}(\max(b, d, e) - \max(b, e)).\text{sg}(\min(a, f.\text{sg}(f - d)) - \min(a, c, f))$$

   2.1. If $d \le \max(b, e)$, then,
        $$A = 1 - 0 = 1.$$
   2.2. If $d > \max(b, e)$, then,
        $$A = \max(b, e) - \min(a, f.\text{sg}(f - d)).\text{sg}(\min(a, f.\text{sg}(f - d)) - \min(a, c, f))$$
        2.2.1. If $f \le d$, then,
               $$A = \max(b, e) - \min(a, 0) = \max(b, e) \ge 0$$

2.2.2. If $f > d$, then,
$$A = \max(b, e) - \min(a, f).\mathrm{sg}(\min(a, f) - \min(a, c, f))$$
 2.2.2.1. If $c \geq \max(a, f)$, then,
$$A = \max(b, e) - 0 \geq 0;$$
 2.2.2.2. If $c < \max(a, f)$, then,
$$A = \max(b, e) - \min(a, f)$$
    2.2.2.2.1. If $a \leq \max(b, e)$, form $f > d > \max(b, e)$, it follows
    that $f > a$ and $A = \max(b, e) - a \geq 0$.
    2.2.2.2.2. If $a > \max(b, e)$, form $f > d > \max(b, e)$, it follows
    that $\min(a, f) > \max(b, e)$

and, therefore, $A < 0$, i.e., the expression

$$((p \wedge q) \supset r) \to (p \supset (q \to r))$$

is valid in all cases without the last one. Hence, it is not an IFT.

On the other hand, the following form of it is valid:

AxiomVW6$'$. $((p \wedge q) \supset r) \to (p \supset (q \supset r))$ because
$V(((p \wedge q) \supset r) \to (p \supset (q \supset r)))$
$= \langle \min(a, c), \max(b, d) \rangle \supset \langle e, f \rangle) \to (\langle a, b \rangle \supset \langle \max(d, e), \min(c, f) \rangle)$
$= \langle \max(b, d, e), \min(a, c, f) \rangle \to \langle \max(b, d, e)), \min(a, c, f) \rangle$
$= \langle 1 - (1 - \max(b, d, e)).\mathrm{sg}(\max(b, d, e) - \max(b, d, e)), \min(a, c, f)$
$\quad .\mathrm{sg}(\max(b, d, e) - \max(b, d, e)).\mathrm{sg}(\min(a, c, f) - \min(a, c, f)) \rangle$
$= \langle 1, 0 \rangle,$
i.e.,
$$((p \wedge q) \supset r) \to (p \supset (q \supset r))$$

is an IFT.

For **Axiom VW7** we obtain:
$V(((p \supset q) \wedge (q \supset p)) \to ((p \supset r) \equiv (q \supset r)))$
$= (\langle \max(b, c), \min(a, d) \rangle \wedge \langle \max(a, d), \min(b, c) \rangle) \to (((\langle \max(b, e), \min(a, f) \rangle$
$\quad \supset \langle \max(d, e), \min(c, f) \rangle \wedge (\langle \max(d, e), \min(c, f) \rangle \supset \langle \max(b, e),$
$\quad \min(a, f) \rangle)))$
$= \langle \min(\max(b, c), \max(a, d)), \max(\min(a, d), \min(b, c)) \rangle$
$\quad \to (\langle \max(\min(a, f), d, e), \min(c, f, \max(b, e)) \rangle$
$\quad \wedge \langle \max(\min(c, f), b, e), \min(a, f, \max(d, e)) \rangle))$
$= \langle \min(\max(b, c), \max(a, d)), \max(\min(a, d), \min(b, c)) \rangle \to$
$\quad \langle \min(\max(\min(a, f), d, e), \max(\min(c, f), b, e)), \max(\min(a, f, \max(d, e)),$
$\quad \min(c, f, \max(b, e))) \rangle)$
$= \langle \max(\min(a, d), \min(b, c), \min(\max(\min(a, f), d, e), \max(\min(c, f), b, e))),$
$\quad \min(\max(a, d), \max(b, c), \max(\min(a, f, \max(d, e)), \min(c, f, \max(b, e)))) \rangle.$
 Let
$A \equiv \max(\min(a, d), \min(b, c), \min(\max(\min(a, f), d, e), \max(\min(c, f), b, e)))$

$- \min(\max(a, d), \max(b, c), \max(\min(a, f, \max(d, e)),$
$\min(c, f, \max(b, e))))$.
From the equalities:

$$\min(\max(a, x), \max(b, x)) = \max(x, \min(a, b))$$

$$\max(\min(a, x), \min(b, x)) = \min(x, \max(a, b))$$

it follows that:
$A = \max(\min(a, d), \min(b, c), \max(e, \min(\min(a, f), d, \min(c, f), b)))$
$\quad \min(\max(a, d), \max(b, c), \min(f, \max(a, \max(d, e), c, \max(b, e))))$
$= \max(\min(a, d), \min(b, c), \max(e, \min(a, b, c, d, f))) - \min(\max(a, d),$
$\quad \max(b, c), \min(f, \max(a, b, c, d, e)))$

1. If $e \geq f$, then
   $A \geq \max(e, \min(a, b, c, d, f)) - \min(f, \max(a, b, c, d, e)) \geq e - f \geq 0$;

2. If $e < f$, then, e.g., for $a = b = f > c = d = e$ we obtain:
   $A = \max(c, d, e) - \min(a, b, f) < 0$,

i.e., in this case Axiom VW7 is not an IFT.

On the other hand, the following form of **Axiom VW7** is valid:
$V(((p \to q) \wedge (q \to p)) \supset ((p \to r) \equiv (q \to r)))$
$= (\langle 1 - (1 - c).\mathrm{sg}(a - c), d.\mathrm{sg}(a - c).\mathrm{sg}(d - b)\rangle \wedge \langle 1 - (1 - a).\mathrm{sg}(c - a),$
$\quad b.\mathrm{sg}(c - a).\mathrm{sg}(b - d)\rangle) \supset (((\langle 1 - (1 - e).\mathrm{sg}(a - e), f.\mathrm{sg}(a - e).\mathrm{sg}(f - b)\rangle$
$\quad \supset \langle 1 - (1 - e).\mathrm{sg}(c - e), f.\mathrm{sg}(c - e).\mathrm{sg}(f - d)\rangle) \wedge (\langle 1 - (1 - e).\mathrm{sg}(c - e),$
$\quad f.\mathrm{sg}(c - e).\mathrm{sg}(f - d)\rangle \supset \langle 1 - (1 - e).\mathrm{sg}(a - e), f.\mathrm{sg}(a - e).\mathrm{sg}(f - b)\rangle)))$
$= \langle \min(1 - (1 - c).\mathrm{sg}(a - c), 1 - (1 - a).\mathrm{sg}(c - a)), \max(d.\mathrm{sg}(a - c).\mathrm{sg}(d - b),$
$\quad b.\mathrm{sg}(c - a).\mathrm{sg}(b - d)\rangle \supset ((\langle \max(f.\mathrm{sg}(a - e).\mathrm{sg}(f - b), 1 - (1 - e).\mathrm{sg}(c - e)),$
$\quad \min(1 - (1 - e).\mathrm{sg}(a - e), f.\mathrm{sg}(c - e).\mathrm{sg}(f - d))\rangle$
$= \langle \max(f.\mathrm{sg}(c - e).\mathrm{sg}(f - d)), 1 - (1 - e).\mathrm{sg}(a - e)),$
$\quad \min(1 - (1 - e).\mathrm{sg}(c - e), f.\mathrm{sg}(a - e).\mathrm{sg}(f - b))\rangle)$
$= \langle \min(1 - (1 - c).\mathrm{sg}(a - c), 1 - (1 - a).\mathrm{sg}(c - a)), \max(d.\mathrm{sg}(a - c).\mathrm{sg}(d - b),$
$\quad b.\mathrm{sg}(c - a).\mathrm{sg}(b - d)\rangle \supset \langle \min(\max(f.\mathrm{sg}(a - e).\mathrm{sg}(f - b), 1 - (1 - e).\mathrm{sg}(c - e)),$
$\quad \max(f.\mathrm{sg}(c - e).\mathrm{sg}(f - d), 1 - (1 - e).\mathrm{sg}(a - e))), \max(\min(1 - (1 - e).\mathrm{sg}(a - e),$
$\quad f.\mathrm{sg}(c - e).\mathrm{sg}(f - d)), \min(1 - (1 - e).\mathrm{sg}(c - e), f.\mathrm{sg}(a - e).\mathrm{sg}(f - b)))\rangle$
$= \langle \max(d.\mathrm{sg}(a - c).\mathrm{sg}(d - b), b.\mathrm{sg}(c - a).\mathrm{sg}(b - d), \min(\max(f.\mathrm{sg}(a - e).\mathrm{sg}(f - b),$
$\quad 1 - (1 - e).\mathrm{sg}(c - e)), \max(f.\mathrm{sg}(c - e).\mathrm{sg}(f - d), 1 - (1 - e).\mathrm{sg}(a - e)))),$
$\quad \min(1 - (1 - c).\mathrm{sg}(a - c), 1 - (1 - a).\mathrm{sg}(c - a), \max(\min(1 - (1 - e)$
$\quad .\mathrm{sg}(a - e), f.\mathrm{sg}(c - e).\mathrm{sg}(f - d)), \min(1 - (1 - e).\mathrm{sg}(c - e),$
$\quad f.\mathrm{sg}(a - e).\mathrm{sg}(f - b))))\rangle$.
Let
$A \equiv \max(d.\mathrm{sg}(a - c).\mathrm{sg}(d - b), b.\mathrm{sg}(c - a).\mathrm{sg}(b - d), \min(\max(f.\mathrm{sg}(a - e)$
$\quad .\mathrm{sg}(f - b), 1 - (1 - e).\mathrm{sg}(c - e)), \max(f.\mathrm{sg}(c - e).\mathrm{sg}(f - d), 1 - (1 - e)$
$\quad .\mathrm{sg}(a - e)))) - \min(1 - (1 - c).\mathrm{sg}(a - c), 1 - (1 - a).\mathrm{sg}(c - a),$

$\max(\min(1-(1-e).\mathrm{sg}(a-e), f.\mathrm{sg}(c-e).\mathrm{sg}(f-d)), \min(1-(1-e).\mathrm{sg}(c-e),$
$f.\mathrm{sg}(a-e).\mathrm{sg}(f-b)))$.

1. Let $a \leq c$, then:
   $A = \max(0, b.\mathrm{sg}(b-d), \min(\max(f.\mathrm{sg}(a-e).\mathrm{sg}(f-b), 1-(1-e).\mathrm{sg}(c-e)),$
   $\max(f.\mathrm{sg}(c-e).\mathrm{sg}(f-d), 1-(1-e).\mathrm{sg}(a-e))))-\min(1, 1-(1-a).\mathrm{sg}(c-a),$
   $\max(\min(1-(1-e).\mathrm{sg}(a-e), f.\mathrm{sg}(c-e).\mathrm{sg}(f-d)), \min(1-(1-e).\mathrm{sg}(c-e),$
   $f.\mathrm{sg}(a-e).\mathrm{sg}(f-b))))$
   $= \max(b.\mathrm{sg}(b-d), \min(\max(f.\mathrm{sg}(a-e).\mathrm{sg}(f-b), 1-(1-e).\mathrm{sg}(c-e)),$
   $\max(f.\mathrm{sg}(c-e).\mathrm{sg}(f-d), 1-(1-e).\mathrm{sg}(a-e))))-\min(1-(1-a).\mathrm{sg}(c-a),$
   $\max(\min(1-(1-e).\mathrm{sg}(a-e), f.\mathrm{sg}(c-e).\mathrm{sg}(f-d)), \min(1-(1-e).\mathrm{sg}(c-e),$
   $f.\mathrm{sg}(a-e).\mathrm{sg}(f-b))))$.

   1.1. If $a \leq e$, then
      $A = \min(\max(0, 1-(1-e).\mathrm{sg}(c-e)), \max(f.\mathrm{sg}(c-e).\mathrm{sg}(f-d), 1))$
      $- \min(1-(1-a).\mathrm{sg}(c-a), \max(\min(1, f.\mathrm{sg}(c-e).\mathrm{sg}(f-d)),$
      $\min(1-(1-e).\mathrm{sg}(c-e), 0)))$
      $= \min(1-(1-e).\mathrm{sg}(c-e), 1)-\min(1-(1-a).\mathrm{sg}(c-a),$
      $\max(f.\mathrm{sg}(c-e).\mathrm{sg}(f-d), 0))$
      $= 1-(1-e).\mathrm{sg}(c-e)-\min(1-(1-a).\mathrm{sg}(c-a), f.\mathrm{sg}(c-e).\mathrm{sg}(f-d))$
      1.1.1. If $c \leq e$, then
         $A = 1 - \min(1-(1-a).\mathrm{sg}(c-a), 0) = 1 - 0 = 1;$
      1.1.2. If $c > e$, then, $c > a$ and $A = e - \min(a, f.\mathrm{sg}(f-d)) \geq e - a \geq 0;$
   1.2. If $a > e$, then, $c > e$ and
      $A = \max(b.\mathrm{sg}(b-d), \min(\max(f.\mathrm{sg}(f-b), e), \max(f.\mathrm{sg}(f-d), 1-e)))$
      $- \min(1-(1-a).\mathrm{sg}(c-a), \max(\min(1-e, f.\mathrm{sg}(f-d)), \min(1-$
      $e, f.\mathrm{sg}(f-b)))$
      1.2.1. If $f \leq b$, then,
         $A = \max(b.\mathrm{sg}(b-d), \min(\max(0, e), \max(f.\mathrm{sg}(f-d), 1-e)))-$
         $\min(1-(1-a)$
         $.\mathrm{sg}(c-a), \max(\min(1-e, f.\mathrm{sg}(f-d)), \min(1-e, 0))$
         $= \max(b.\mathrm{sg}(b-d), \min(e, \max(f.\mathrm{sg}(f-d), 1-e)))-\min(1-(1-$
         $a).\mathrm{sg}(c-a),$
         $\max(\min(1-e, f.\mathrm{sg}(f-d)), 0))$
         $= \max(b.\mathrm{sg}(b-d), \min(e, \max(f.\mathrm{sg}(f-d), 1-e)))-\min(1-(1-$
         $a).\mathrm{sg}(c-a),$
         $1-e, f.\mathrm{sg}(f-d))$
         1.2.1.1. If $f \leq d$, then,
            $A = \max(b.\mathrm{sg}(b-d), \min(e, \max(0, 1-e)))-\min(1-(1-$
            $a).\mathrm{sg}(c-a), 1-e, 0)$
            $= \max(b.\mathrm{sg}(b-d), \min(e, 1-e))-0 \geq 0;$
         1.2.1.2. If $f > d$, then, from $e + f \leq 1$
            $A = \max(b, \min(e, \max(f, 1-e)))-\min(1-(1-a).\mathrm{sg}(c-a), 1-$
            $e, f)$

$$= \max(b, \min(e, 1 - e)) - \min(1 - (1 - a).\text{sg}(c - a), f)$$

1.2.1.2.1. If $c = a$ (below we assume that $c \geq a$) then,

$$A = \max(b, \min(e, 1 - e)) - \min(1, f)$$
$$= \max(b, \min(e, 1 - e)) - f \geq b - f \geq 0;$$

1.2.1.2.2. If $c > a$, then,

$$A = \max(b, \min(e, 1 - e)) - \min(a, f) \geq b - f \geq 0;$$

1.2.2. If $f > b$, then,

$$A = \max(b.\text{sg}(b - d), \min(\max(f, e), \max(f.\text{sg}(f - d), 1 - e)))$$
$$- \min(1 - (1 - a).\text{sg}(c - a), \max(\min(1 - e, f.\text{sg}(f - d)),$$
$$\min(1 - e, f)))$$

1.2.2.1. If $f \leq d$, then, $d > b$ and

$$A = \max(0, \min(\max(f, e), \max(0, 1 - e))) - \min(1 - (1 - a).\text{sg}(c - a), \max(\min(1 - e, 0), \min(1 - e, f)))$$
$$= \min(\max(f, e), 1 - e) - \min(1 - (1 - a).\text{sg}(c - a), \max(0, \min(1 - e, f)))$$
$$= \min(\max(f, e), 1 - e) - \min(1 - (1 - a).\text{sg}(c - a), f)$$
$$\geq \min(\max(f, e), 1 - e) - f \geq 0,$$

because as $\max(f, e) \geq f$, as well as $1 - e \geq f$;

1.2.2.2. If $f > d$, then,

$$A = \max(b.\text{sg}(b - d), \min(\max(f, e), \max(f, 1 - e))) - \min(1 - (1 - a).\text{sg}(c - a),$$
$$\max(\min(1 - e, f), \min(1 - e, f)))$$
$$= \max(b.\text{sg}(b - d), \min(\max(f, e), 1 - e)) - \min(1 - (1 - a).\text{sg}(c - a), \max(f, f))$$
$$= \max(b.\text{sg}(b - d), \min(\max(f, e), 1 - e)) - \min(1 - (1 - a).\text{sg}(c - a), f)$$
$$\geq \min(\max(f, e), 1 - e) - f \geq 0;$$

2. Let $a > c$, then

$$A = \max(d.\text{sg}(d - b), 0, \min(\max(f.\text{sg}(a - e).\text{sg}(f - b), 1 - (1 - e).\text{sg}(c - e)),$$
$$\max(f.\text{sg}(c - e).\text{sg}(f - d), 1 - (1 - e).\text{sg}(a - e)))) - \min(c, 1,$$
$$\max(\min(1 - (1 - e).\text{sg}(a - e), f.\text{sg}(c - e).\text{sg}(f - d)), \min(1 - (1 - e).\text{sg}(c - e), f.\text{sg}(a - e).\text{sg}(f - b))))$$
$$= \max(d.\text{sg}(d - b), \min(\max(f.\text{sg}(a - e).\text{sg}(f - b), 1 - (1 - e).\text{sg}(c - e)),$$
$$\max(f.\text{sg}(c - e).\text{sg}(f - d), 1 - (1 - e).\text{sg}(a - e)))) - \min(c, \max(\min(1 - (1 - e).\text{sg}(a - e), f.\text{sg}(c - e).\text{sg}(f - d)), \min(1 - (1 - e).\text{sg}(c - e), f.\text{sg}(a - e).\text{sg}(f - b))))$$

2.1. If $a \leq e$, then, $e > c$ and

$$A = \max(d.\text{sg}(d - b), \min(\max(0, 1), \max(0, 1)))) - \min(c,$$
$$\max(\min(e, 0), \min(1, 0)))$$
$$= \max(d.\text{sg}(d - b), \min(1, 1)) - \min(c, \max(0, 0))$$
$$= \max(d.\text{sg}(d - b), 1) - \min(c, 0)$$
$$= 1 - 0 = 1;$$

2.2. If $a > e$, then,

$$A = \max(d.\text{sg}(d - b), \min(\max(f.\text{sg}(f - b), 1 - (1 - e).\text{sg}(c - e)),$$

$$\max(f.\mathrm{sg}(c - e).\mathrm{sg}(f - d), e))) - \min(c, \max(\min(e,$$
$$f.\mathrm{sg}(c - e).\mathrm{sg}(f - d)), \min(1 - (1 - e).\mathrm{sg}(c - e), f.\mathrm{sg}(f - b))))$$

2.2.1. If $f \leq b$, then,

$A = \max(d.\mathrm{sg}(d - b), \min(\max(0, 1 - (1 - e).\mathrm{sg}(c - e)), \max(f.\mathrm{sg}(c - e).\mathrm{sg}(f - d),$

$\quad e))) - \min(c, \max(\min(e, f.\mathrm{sg}(c - e).\mathrm{sg}(f - d)), \min(1 - (1 - e).\mathrm{sg}(c - e), 0)))$

$= \max(d.\mathrm{sg}(d - b), \min(1 - (1 - e).\mathrm{sg}(c - e), \max(f.\mathrm{sg}(c - e).\mathrm{sg}(f - d), e))) - \min(c,$

$\quad \max(\min(e, f.\mathrm{sg}(c - e).\mathrm{sg}(f - d)), 0))$

$= \max(d.\mathrm{sg}(d - b), \min(1 - (1 - e).\mathrm{sg}(c - e), \max(f.\mathrm{sg}(c - e).\mathrm{sg}(f - d), e)))$

$\quad - \min(c, \min(e, f.\mathrm{sg}(c - e).\mathrm{sg}(f - d)))$

2.2.1.1. If $f \leq d$, then,

$A = \max(d.\mathrm{sg}(d - b), \min(1 - (1 - e).\mathrm{sg}(c - e), \max(0, e))) - \min(c, \min(e, 0))$

$= \max(d.\mathrm{sg}(d - b), \min(1 - (1 - e).\mathrm{sg}(c - e), e)) - \min(c, 0)$

$= \max(d.\mathrm{sg}(d - b), \min(1 - (1 - e).\mathrm{sg}(c - e), e)) - 0 \geq 0;$

2.2.1.2. If $f > d$, then, $b > d$ and

$A = \max(0, \min(1 - (1 - e).\mathrm{sg}(c - e), \max(f.\mathrm{sg}(c - e), e)))$

$\quad - \min(c, \min(e, f.\mathrm{sg}(c - e)))$

$= \min(1 - (1 - e).\mathrm{sg}(c - e), \max(f.\mathrm{sg}(c - e), e))) - \min(c, \min$

$(e, f.\mathrm{sg}(c - e)))$

2.2.1.2.1. If $c \leq e$, then,

$A = \min(1, \max(0, e)) - \min(c, \min(e, 0))$

$= \min(1, e) - \min(c, 0) = e - 0 \geq 0;$

2.2.1.2.2. If $c > e$, then,

$A = \min(e, \max(f, e)) - \min(c, \min(e, f))$

$= e - \min(c, e, f) \geq 0;$

2.2.2. If $f > b$, then,

$A = \max(d.\mathrm{sg}(d - b), \min(\max(f, 1 - (1 - e).\mathrm{sg}(c - e)), \max(f.\mathrm{sg}(c - e).\mathrm{sg}(f - d), e))) - \min(c, \max(\min(e, f.\mathrm{sg}(c - e).\mathrm{sg}(f - d)), \min(1 - (1 - e).\mathrm{sg}(c - e), f)))$

2.2.2.1. If $f \leq d$, then,

$A = \max(d.\mathrm{sg}(d - b), \min(\max(f, 1 - (1 - e).\mathrm{sg}(c - e)), \max(0, e)))$

$\quad - \min(c, \max(\min(e, 0), \min(1 - (1 - e).\mathrm{sg}(c - e), f)))$

$= \max(d.\mathrm{sg}(d - b), \min(\max(f, 1 - (1 - e).\mathrm{sg}(c - e)), e))$

$\quad - \min(c, \max(0, \min(1 - (1 - e).\mathrm{sg}(c - e), f)))$

$= \max(d.\mathrm{sg}(d - b), \min(\max(f, 1 - (1 - e).\mathrm{sg}(c - e)), e))$

$\quad - \min(c, 1 - (1 - e).\mathrm{sg}(c - e), f)$

2.2.2.1.1. If $c \leq e$, then,

$A = \max(d.\mathrm{sg}(d - b), \min(\max(f, 1), e)) - \min(c, 1, f)$

$$= \max(d.\mathrm{sg}(d - b), \min(1, e)) - \min(c, f)$$
$$= \max(d.\mathrm{sg}(d - b), e) - \min(c, f) \geq e - c \geq 0;$$

2.2.2.1.2. If $c > e$, then,
$$A = \max(d.\mathrm{sg}(d-b), \min(\max(f, e), e)) - \min(c, e, f)$$
$$= \max(d.\mathrm{sg}(d-b), e) - \min(c, e, f) \geq e - \min(c, e, f)$$
$$\geq 0;$$

2.2.2.2. If $f > d$, then,
$$A = \max(d.\mathrm{sg}(d - b), \min(\max(f, 1 - (1 - e).\mathrm{sg}(c - e)),$$
$$\max(f.\mathrm{sg}(c - e), e))) - \min(c, \max(\min(e, f.\mathrm{sg}(c - e)),$$
$$\min(1 - (1 - e).\mathrm{sg}(c - e), f)))$$

2.2.2.2.1. If $c \leq e$, then,
$$A = \max(d.\mathrm{sg}(d - b), \min(\max(f, 1), \max(0, e)))$$
$$- \min(c, \max(\min(e, 0), \min(1, f)))$$
$$= \max(d.\mathrm{sg}(d - b), \min(1, e)) - \min(c, \max(0, f))$$
$$= \max(d.\mathrm{sg}(d - b), e) - \min(c, f) \geq e - c \geq 0;$$

2.2.2.2.2. If $c > e$, then,
$$A = \max(d.\mathrm{sg}(d - b), \min(\max(f, e), \max(f, e)))$$
$$- \min(c, \max(\min(e, f), \min(e, f)))$$
$$= \max(d.\mathrm{sg}(d - b), \max(f, e)) - \min(c, \min(e, f))$$
$$\geq \max(f, e) - \min(e, f) \geq 0,$$

i.e., in this case Axiom VW7$'$ is an IFT. $\square$

So, we checked the validity of original or modified Conditional Logic Axioms for pair $(\to_4, \to_{11})$.

Open Problem 4. Which pairs of implications $(\to_i, \to_j)$ for $1 \leq i, j \leq 185$ satisfy these axioms in original or modified forms?

For example, we mention that the original Axioms are valid only for pairs $(\to_{20}, \to_{20})$, $(\to_{23}, \to_{23})$, $(\to_{74}, \to_{74})$ and $(\to_{77}, \to_{77})$, i.e., for the case, when the implications $\supset$ and $\to$ in the seven axioms coincide. If we change the condition for the Axioms to be true with condition for these axioms to be IFTs, then, these pairs with equal components are generated by implications $\to_{20}$, $\to_{23}$, $\to_{27}$, $\to_{29}$, $\to_{74}$, $\to_{77}$, $\to_{81}$, $\to_{101}$, $\to_{102}$, $\to_{111}$, $\ldots$, $\to_{113}$, $\to_{118}$, $\to_{126}$, $\to_{128}$, $\to_{167}$, $\to_{169}$.

1.6   De Morgan Laws and Law for Excluded Middle

Following and extending [59], first, we give the Law for Excluded Middle (LEM) in the forms:
$$V(A \vee \neg A) = \langle 1, 0 \rangle$$

(standard tautology) and

$$V(A \vee \neg A) = \langle p, q \rangle,$$

(IFT), where $1 \geq p \geq q \geq 0$ and $p + q \leq 1$.

Second, we give the Modified Law for Excluded Middle (MLEM) in the forms:

$$\neg\neg\langle a, b \rangle \vee \neg\langle a, b \rangle = \langle 1, 0 \rangle$$

(standard tautology) and

$$\neg\neg\langle a, b \rangle \vee \neg\langle a, b \rangle = \langle p, q \rangle,$$

(IFT), where $1 \geq p \geq q \geq 0$ and $p + q \leq 1$ and $\vee$ is the disjunction from Sect. 1.1.

Theorem 1.6.1 *Only negation $\neg_{13}$ satisfies the LEM in the tautological form.*

Proof Let $V(A) = \langle a, b \rangle$ and $a, b, a + b \in [0, 1]$. Then,

$$V(x \vee \neg_{13}x) = \langle a, b \rangle \vee \langle \text{sg}(1 - a), \overline{\text{sg}}(1 - a) \rangle$$

$$= \langle \max(a, \text{sg}(1 - a)), \min(b, \overline{\text{sg}}(1 - a)) \rangle.$$

Now, we see that

$$\max(a, \text{sg}(1 - a)) = \begin{cases} \max(1, \text{sg}(0)), & \text{if } a = 1 \\ \max(a, \text{sg}(1 - a)), & \text{if } a < 1 \end{cases}$$

$$= \begin{cases} \max(1, 0), & \text{if } a = 1 \\ \max(a, 1), & \text{if } a < 1 \end{cases} = 1$$

and

$$\min(b, \overline{\text{sg}}(1 - a)) = \begin{cases} \min(0, \overline{\text{sg}}(0)), & \text{if } a = 1 \\ \min(0, \text{sg}(1 - a)), & \text{if } a < 1 \end{cases}$$

$$= \begin{cases} \min(0, 1), & \text{if } a = 1 \\ \min(0, 1), & \text{if } a < 1 \end{cases} = 0.$$

Therefore,

$$V(x \vee \neg_{13}x) = \langle 1, 0 \rangle,$$

i.e., $x \vee \neg_{13}x$ is a tautology.   □

Theorem 1.6.2 *Only negations $\neg_2$, $\neg_5$, $\neg_9$, $\neg_{11}$, $\neg_{13}$, $\neg_{16}$ satisfy the MLEM in the tautological form.*

Theorem 1.6.3 *Negations $\neg_1$, $\neg_3$, $\neg_4$, $\neg_7$, ..., $\neg_9$, $\neg_{11}$, ..., $\neg_{34}$, $\neg_{40}$, $\neg_{43}$, $\neg_{48}$, $\neg_{49}$, $\neg_{52}$, $\neg_{53}$ satisfy the LEM in the IFT form.*

Theorem 1.6.4 *Negations* $\neg_1, \ldots, \neg_9, \neg_{11}, \ldots, \neg_{34}, \neg_{43}, \neg_{48}, \neg_{49}, \neg_{52}, \neg_{53}$ *satisfy the MLEM in the IFT form.*

The checks of the last three assertions are similar to the proof of Theorem 1.6.1. By this reason we will prove only Theorems 1.6.2 and 1.6.4 for the case of negation $\neg_5$.

$$\neg_5 \neg_5 \langle a, b \rangle \vee \neg_5 \langle a, b \rangle$$

$$= \langle 1 - \mathrm{sg}(a) - \overline{\mathrm{sg}}(a).\mathrm{sg}(1-b), \mathrm{sg}(1-b) \rangle \vee \langle 1 - \mathrm{sg}(1 - \mathrm{sg}(a) - \overline{\mathrm{sg}}(a).\mathrm{sg}(1-b))$$

$$-\overline{\mathrm{sg}}(1 - \mathrm{sg}(a) - \overline{\mathrm{sg}}(a).\mathrm{sg}(1-b)).\mathrm{sg}(1 - \mathrm{sg}(1-b)), \mathrm{sg}(1 - \mathrm{sg}(1-b)) \rangle$$

$$= \langle \max(1 - \mathrm{sg}(a) - \overline{\mathrm{sg}}(a).\mathrm{sg}(1-b), 1 - \mathrm{sg}(1 - \mathrm{sg}(a) - \overline{\mathrm{sg}}(a).\mathrm{sg}(1-b))$$

$$-\overline{\mathrm{sg}}(1-\mathrm{sg}(a)-\overline{\mathrm{sg}}(a).\mathrm{sg}(1-b)).\mathrm{sg}(1-\mathrm{sg}(1-b))), \min(\mathrm{sg}(1-b), \mathrm{sg}(1-\mathrm{sg}(1-b))) \rangle.$$

Let

$$X \equiv \max(1 - \mathrm{sg}(a) - \overline{\mathrm{sg}}(a).\mathrm{sg}(1-b), 1 - \mathrm{sg}(1 - \mathrm{sg}(a) - \overline{\mathrm{sg}}(a).\mathrm{sg}(1-b))$$

$$-\overline{\mathrm{sg}}(1 - \mathrm{sg}(a) - \overline{\mathrm{sg}}(a).\mathrm{sg}(1-b)).\mathrm{sg}(1 - \mathrm{sg}(1-b)))$$

$$- \min(\mathrm{sg}(1-b), \mathrm{sg}(1 - \mathrm{sg}(1-b))).$$

Let $a = 0$. Then, $\mathrm{sg}(a) = 0$, $\overline{\mathrm{sg}}(a) = 1$ and

$$X = \max(1 - \mathrm{sg}(1-b), 1 - \mathrm{sg}(1 - \mathrm{sg}(1-b)) - \overline{\mathrm{sg}}(1 - \mathrm{sg}(1-b)).\mathrm{sg}(1 - \mathrm{sg}(1-b)))$$

$$- \min(\mathrm{sg}(1-b), \mathrm{sg}(1 - \mathrm{sg}(1-b))).$$

If $b = 1$, then, $\mathrm{sg}(1-b) = 0$ and

$$X = \max(1, 1 - \mathrm{sg}(1) - \overline{\mathrm{sg}}(1).\mathrm{sg}(1)) - \min(0, \mathrm{sg}(1)) = \max(1, 0) - \min(0, 1) = 1.$$

If $b < 1$, then, $\mathrm{sg}(1-b) = 1$ and

$$X = \max(1 - 1, 1 - \mathrm{sg}(1 - 1) - \overline{\mathrm{sg}}(1 - 1).\mathrm{sg}(1 - 1)) - \min(1, \mathrm{sg}(1 - 1))$$

$$= \max(0, 1) - \min(1, 0) = 1.$$

Let $a > 0$. Then, $\mathrm{sg}(a) = 1$, $\overline{\mathrm{sg}}(a) = 0$, $\mathrm{sg}(1-b) = 1$ and

$$X \equiv \max(1 - 1, 1 - \mathrm{sg}(1 - 1) - \overline{\mathrm{sg}}(1 - 1).\mathrm{sg}(1 - 1)) - \min(1, \mathrm{sg}(1 - 1))$$

$$= \max(0, 1) - \min(1, 0) = 1.$$

Therefore, negation $\neg_5$ satisfies the Modified LEM in the IFT-form. On the other hand, in all cases the evaluation of the expression is equal to $\langle 1, 0 \rangle$, i.e., this negation satisfies the Modified LEM in the tautological form.   □

Third, following and extending [76], we study which negations satisfy De Morgan Laws (DMLs). Usually, they have the forms:

$$\neg x \wedge \neg y = \neg(x \vee y),$$

$$\neg x \vee \neg y = \neg(x \wedge y),$$

where $\wedge$ and $\vee$ are the conjunction and disjunction from Sect. 1.1, see (1.1.4) and (1.1.5).

Theorem 1.6.5 *For every two formulas A and B:*

$$\neg_i A \wedge \neg_i B = \neg_i(A \vee B),$$

$$\neg_i A \vee \neg_i B = \neg_i(A \wedge B)$$

for $i = 1, 2, 4, \ldots, 11, 13, \ldots, 17, 20, 23, 35, \ldots, 51, 53.$
We shall illustrate only the fact that the DMLs are not valid for $i = 3$. For example, if $a = b = 0.5, c = 0.1, d = 0$, then

$$V(\neg_3 A \wedge \neg_3 B) = 0.5,$$

$$V(\neg_3(A \vee B)) = 0.25.$$

The above mentioned change of the Law for Excluded Middle inspired the idea to study the validity of De Morgan's Laws, which the classical negation $\neg$ (here it is negation $\neg_1$) satisfies. Indeed, it can be easily proved that the expressions

$$\neg_1(\neg_1 A \vee \neg_1 B) = A \wedge B$$

and

$$\neg_1(\neg_1 A \wedge \neg_1 B) = A \vee B$$

are IFTs, while the other negations do not satisfy these equalities. For them the following assertion is valid.

Theorem 1.6.6 *For every two formulas A and B, it holds that*

$$\neg_i(\neg_i A \vee \neg_i B) = \neg_i \neg_i A \wedge \neg_i \neg_i B,$$

$$\neg_i(\neg_i A \wedge \neg_i B) = \neg_i \neg_i A \vee \neg_i \neg_i B$$

for $i = 1, 2, 4, \ldots, 11, 13, \ldots, 17, 19, 20, 23, 25, 35, \ldots, 51, 53.$

1.7   New Intuitionistic Fuzzy Conjunctions and Disjunctions

From classical logic, it is well-known that for any two formulas A and B:

$$A \vee B = \neg A \rightarrow B, \tag{1.7.1}$$

$$A \wedge B = \neg (A \rightarrow \neg B). \tag{1.7.2}$$

Therefore, having the above intuitionistic fuzzy implications and negations, we can construct 185 disjunctions and 185 conjunctions. In Sect. 1.2 we saw that implication (1.1.6), which is denoted there by $\rightarrow_4$, and negation (1.1.3), which in Sect. 1.4 is denoted by $\neg_1$, are connected by equality (1.2.1). So, using formulas (1.7.1) and (1.7.2) we can construct from them a disjunction and a conjunction, and they exactly coincide with these from (1.1.4) and (1.1.5), respectively. All other disjunctions and conjunctions can be constructed in the same manner. For example, if $V(A) = \langle a, b \rangle, V(B) = \langle c, d \rangle, a, b, c, d \in [0, 1], a + b \le 1, c + d \le 1$ and if we use negation $\neg_2$ and its related implication $\rightarrow_2$, we obtain sequentially:

$$V(A \vee_2 B) = \langle a, b \rangle \vee_2 \langle c, d \rangle$$

$$= \neg_2 \langle a, b \rangle \rightarrow_2 \langle c, d \rangle$$

$$= \langle \overline{sg}(a), sg(a) \rangle \rightarrow_2 \langle c, d \rangle$$

$$= \langle \overline{sg}(\overline{sg}(a) - c), d sg(\overline{sg}(a) - c) \rangle$$

and

$$V(A \wedge_2 B) = \langle a, b \rangle \wedge_2 \langle c, d \rangle$$

$$= \neg_2 (\langle a, b \rangle \rightarrow_2 \neg_2 \langle c, d \rangle)$$

$$= \neg_2 (\langle a, b \rangle \rightarrow_2 \langle \overline{sg}(c), sg(c) \rangle)$$

$$= \neg_2 \langle \overline{sg}(a - \overline{sg}(c)), sg(c) sg(a - \overline{sg}(c)) \rangle$$

$$= \langle \overline{sg}(\overline{sg}(a - \overline{sg}(c))), sg(\overline{sg}(a - \overline{sg}(c))) \rangle$$

$$= \langle sg(a - \overline{sg}(c)), \overline{sg}(a - \overline{sg}(c)) \rangle.$$

As we saw in Sect. 1.6, the disjunctions and conjunctions can have two forms. Therefore, formulas (1.7.1) and (1.7.2) can be changed with the new ones:

$$A \vee B = \neg A \rightarrow \neg\neg B, \tag{1.7.3}$$

$$A \wedge B = \neg(\neg\neg A \rightarrow \neg B). \tag{1.7.4}$$

For example, for the above formulas A and B, negation $\neg_2$ and implication $\rightarrow_2$, we obtain sequentially:

$$V(A \vee_2 B) = \neg_2\langle a, b\rangle \vee_2 \neg_2\neg_2\langle c, d\rangle$$

$$= \neg_2\langle a, b\rangle \rightarrow_2 \neg_2\langle \overline{sg}(c), sg(c)\rangle$$

$$= \neg_2\langle a, b\rangle \rightarrow_2 \langle \overline{sg}(\overline{sg}(c)), sg(\overline{sg}(c))\rangle$$

(For every $x \in [0, 1]$:

$$\overline{sg}(\overline{sg}(x)) = \begin{cases} 0, & \text{if } x = 0 \\ 1, & \text{if } x > 0 \end{cases} = sg(x)$$

and

$$sg(\overline{sg}(x)) = \begin{cases} 1, & \text{if } x = 0 \\ 0, & \text{if } x > 0 \end{cases} = \overline{sg}(x).$$

$$= \langle \overline{sg}(a), sg(a)\rangle \rightarrow_2 \langle sg(c), \overline{sg}(c)\rangle$$

$$= \langle \overline{sg}(\overline{sg}(a) - sg(c)), \overline{sg}(c)sg(\overline{sg}(a) - sg(c))\rangle$$

and

$$V(A \wedge_2 B) = \langle a, b\rangle \wedge_2 \langle c, d\rangle$$

$$= \neg_2(\neg_2\neg_2\langle a, b\rangle \rightarrow_2 \neg_2\langle c, d\rangle)$$

$$= \neg_2(\neg_2\langle \overline{sg}(a), sg(a)\rangle \rightarrow_2 \langle \overline{sg}(c), sg(c)\rangle)$$

$$= \neg_2(\langle \overline{sg}(\overline{sg}(a)), sg(\overline{sg}(a))\rangle \rightarrow_2 \langle \overline{sg}(c), sg(c)\rangle)$$

$$= \neg_2(\langle sg(a), \overline{sg}(a)\rangle \rightarrow_2 \langle \overline{sg}(c), sg(c)\rangle)$$

$$= \neg_2\langle \overline{sg}(sg(a) - \overline{sg}(c)), sg(c)sg(sg(a) - \overline{sg}(c))\rangle$$

$$= \langle \overline{sg}(\overline{sg}(sg(a) - \overline{sg}(c))), sg(\overline{sg}(sg(a) - \overline{sg}(c)))\rangle$$

$$= \langle sg(sg(a) - \overline{sg}(c)), \overline{sg}(sg(a) - \overline{sg}(c))\rangle.$$

Having in mind that a part of the disjunctions and conjunctions are generated by the classical negation $\neg_1$, we call all these *"semiclassical disjunctions and conjunctions"*, excluding only the disjunction $\vee_4$ and conjunction $\wedge_4$, i.e., the disjunction $\vee$

and conjunction $\wedge$ from Sect. 1.1, that we call *"classical disjunctions and conjunctions"*. We call the rest disjunctions and conjunctions *"non-classical disjunctions and conjunctions"*.

Therefore, formulas (1.7.1)–(1.7.4) must be rewritten to

$$A \vee_{i,1} B = \neg_{\varphi(i)} A \rightarrow_i B, \tag{1.7.5}$$

$$A \wedge_{i,1} B = \neg_{\varphi(i)} (A \rightarrow_i \neg_{\varphi(i)} B), \tag{1.7.6}$$

$$A \vee_{i,2} B = \neg_{\varphi(i)} A \rightarrow_i \neg_{\varphi(i)} \neg_{\varphi(i)} B, \tag{1.7.7}$$

$$A \wedge_{i,2} B = \neg_{\varphi(i)} (\neg_{\varphi(i)} \neg_{\varphi(i)} A \rightarrow_i \neg_{\varphi(i)} B), \tag{1.7.8}$$

where $\varphi(i)$ is the number of the negation that corresponds to the i-th implication (cf. Table 1.3).

Now, we see a possibility for constructing a third group of disjunctions and conjunctions. They have the forms

$$A \vee_{i,3} B = \neg_1 A \rightarrow_i B, \tag{1.7.9}$$

$$A \wedge_{i,3} B = \neg_1 (A \rightarrow_i \neg_1 B). \tag{1.7.10}$$

Therefore, in all of them only the classical negation $\neg_1$ is used and by this reason, we call them again *"semiclassical disjunctions and conjunctions"*, excluding as above only disjunction $\vee_{4,3}$ and conjunction $\wedge_{4,3}$.

For example, for the above formulas A and B, negation $\neg_2$ and implication $\rightarrow_2$, we obtain sequentially:

$$V(A \vee_{2,3} B) = \neg_1 \langle a, b \rangle \vee_2 \neg_1 \neg_1 \langle c, d \rangle$$

$$= \langle b, a \rangle \vee_2 \langle c, d \rangle$$

$$= \langle \overline{sg}(b - c), d.sg(b - c) \rangle$$

and

$$V(A \wedge_{2,3} B) = \neg_1 (\neg_1 \neg_1 \langle a, b \rangle \rightarrow_2 \neg_1 \langle c, d \rangle)$$

$$= \neg_1 (\langle a, b \rangle \rightarrow_2 \langle d, c \rangle)$$

$$= \neg_1 \langle \overline{sg}(a - d), c.sg(a - d) \rangle$$

$$= \langle c.sg(a - d), \overline{sg}(a - d) \rangle.$$

Three open problems arise here.

Open Problem 5. Construct all new disjunctions and conjunctions. It is interesting to check whether some disjunctions and conjunctions will coincide.

Open Problem 6. Study the behaviour of the new disjunctions and conjunctions. For example, which of them will satisfy De Morgan's Laws and in which form of these Laws?

Open Problem 7. Study the properties of the disjunctions and conjunctions from (1.7.5)–(1.7.10). It is very important to check the validity of the separate axioms – of the intuitionistic logic, of Kolmogorov, of Łukasiewicz and Tarski, of Klir and Yuan, and the other ones, discussed in Sect. 1.5.

We finish this chapter with a remark from the area of group theory.

Here, for the first time, the author described an idea generated by him about 45 year ago, when he was a schoolboy in the secondary school. Only in the last months, after discussions with colleagues, he collected enthusiasm to formulate it (of course, in essentially better form than in the beginning).

Let for a fixed set X, $P(X) = \{Y | Y \subseteq X\}$.

Obviously, if $X \neq \emptyset$ is a fixed set, then, $\langle \mathcal{P}(X), \wedge, X \rangle$ and $\langle \mathcal{P}(X), \vee, \emptyset \rangle$ are commutative monoids, but for every set $A \in \mathcal{P}(X)$ there is no element B such that $A \wedge B = X$ and $A \vee B = \emptyset$. By this reason, $\langle \mathcal{P}(X), \wedge, X \rangle$ and $\langle \mathcal{P}(X), \vee, \emptyset \rangle$ are not (commutative) groups.

We call $\langle M, *, e_*, e_\circ \rangle$ a "multi unitary group" (shortly, μ-group) if and only if

$$(\forall a, b \in M)(a * b \in M); \tag{1.7.11}$$

$$(\forall a, b, c \in M)((a * b) * c = a * (b * c)); \tag{1.7.12}$$

$$(\forall a \in M)(a * e_* = a = e_* * a); \tag{1.7.13}$$

$$(\forall a \in M)(\exists a_\circ \in M)(a * a_\circ = e_\circ = a_\circ * a). \tag{1.7.14}$$

The μ-group is commutative if and only if

$$(\forall a, b \in M)(a * b = b * a). \tag{1.7.15}$$

For example, $\langle P(X), \wedge, X, \emptyset \rangle$ and $\langle P(X), \vee, \emptyset, X \rangle$ are μ-groups.

In the particular case, when $e_* = e_\circ$, the (commutative) μ-group is reduced to a standard (commutative) group.

Two μ-groups MG_1 and MG_2 are dual, if and only if they have the forms

$$MG_1 = \langle M, *, e_*, e_\circ \rangle \quad \text{and} \quad MG_2 = \langle M, \circ, e_\circ, e_* \rangle$$

for some given operations $*$ and $\circ$, and for the unitary elements e_* and $e_\circ$.

For example, $\langle P(X), \wedge, X, \emptyset \rangle$ and $\langle P(X), \vee, \emptyset, X \rangle$ are dual μ-groups.

Formulas (1.7.13) and (1.7.14) can be written in more details, if we like to define left- and right-μ-group. All formulas (1.7.11)–(1.7.15) have analogues in group theory.

In the intuitionistic fuzzy case, we must modify condition (1.7.14) so that for a relation R:

$$(\forall a \in M)(\exists a_\circ \in M)(e_\circ R(a * a_\circ)Re_*). \tag{1.7.16}$$

For example, let $*$ be either the operation $\wedge$ or the operation $\vee$, R be the relation $\subset$, $e_\circ$ be $\emptyset$ and e_* be X. Then, (1.7.16) obtains either the form

$$(\forall a \in M)(\exists a_\circ \in M)(\emptyset \subset (a \wedge a_\circ) \subset X),$$

or

$$(\forall a \in M)(\exists a_\circ \in M)(\emptyset \subset (a \vee a_\circ) \subset X).$$

Now, the following open problem arises.

Open Problem 8. For which numbers i ($1 \leq i \leq 185$), the objects $\langle \mathcal{S}, \vee_i, F, T \rangle$ and $\langle \mathcal{S}, \wedge_i, T, F \rangle$ are (commutative) dual μ-groups, satisfying condition (1.7.16), where $\mathcal{S}, T, F$ are described in Sect. 1.1, relation R is $\leq$ or $\geq$, and the disjunctions and conjunctions are defined by (1.7.5)–(1.7.10)?

References

1. Kleene S. Mathematical logic. New York: Wiley; 1967.
2. Mendelson E. Introduction to mathematical logic. D. Van Nostrand: Princeton; 1964.
3. Shoenfield JR. Mathematical logic. 2nd ed. Natick: A. K. Peters; 2001.
4. http://plato.stanford.edu/entries/generalized-quantifiers/
5. Zadeh L. Fuzzy logic. IEEE Comput. 1988;21(4):83–93.
6. Atanassov K. Two variants of intuitionistic fuzzy propositional calculus. Preprint IM-MFAIS-5-88, Sofia, 1988, Reprinted: Int J Bioautomation. 2016;20(S1):S17–S26.
7. Atanassov K. Intuitionistic fuzzy sets. Heidelberg: Springer; 1999.
8. Atanassov K. On intuitionistic fuzzy logics: results and problems. In: Atanassov K, Baczynski M, Drewniak J, Kacprzyk J, Krawczak M, Szmidt E, Wygralak M, Zadrozny S, editors. Modern approaches in fuzzy sets, intuitionistic fuzzy sets, generalized nets and related topics, vol. 1: Foundations. Warsaw: SRI-PAS; 2014. p. 23–49.
9. Atanassov K. Intuitionistic fuzzy logics as tools for evaluation of Data Mining processes. Knowl-Based Syst. 2015;80:122–130.
10. Atanassov K, Szmidt E, Kacprzyk J. On intuitionistic fuzzy pairs. Notes on Intuitionistic Fuzzy Sets. 2013;19(3):1–13.
11. Atanassov K. On intuitionistic fuzzy sets theory. Berlin: Springer; 2012.
12. Kaufmann A. Introduction a la Theorie des Sour-Ensembles Flous. Paris: Masson; 1977.
13. van Atten M. On Brouwer. Wadsworth: Behnout; 2004.
14. Brouwer LEJ. Collected works, vol. 1. Amsterdam: North Holland; 1975.
15. Brouwer's Cambridge Lectures on Intuitionism (D. Van Dalen, Ed.), Cambridge University Press, Cambridge, 1981.

16. Atanassov K. Geometrical interpretation of the elements of the intuitionistic fuzzy objects, Preprint IM-MFAIS-1-89, Sofia, 1989, Reprinted: Int J Bioautomation. 2016;20(S1):S27–S42.
17. Dworniczak P. A note on the unconscientious experts' evaluations in the intuitionistic fuzzy environment. Notes on Intuitionistic Fuzzy Sets. 2012;18(3):23–29.
18. Dworniczak P. Further remarks about the unconscientious experts evaluations in the intuitionistic fuzzy environment. Notes on Intuitionistic Fuzzy Sets. 2013;19(1):27–31.
19. Atanassova V. Representation of fuzzy and intuitionistic fuzzy data by Radar charts. Notes on Intuitionistic Fuzzy Sets. 2010;16(1):21–26. http://ifigenia.org/wiki/issue:nifs/16/1/21-26
20. Angelova N, Atanassov K. Intuitionistic fuzzy implications and the axioms of intuitionistic logic. In: 9th Conference of the European Society for Fuzzy Logic and Technology (EUSFLAT), 30.06–03.07. Gijon, Spain, 2015. p. 1578–1584.
21. Angelova N, Atanassov K. Intuitionistic fuzzy implications and Klir-Yuans axioms. In: Atanassov KT, Castillo O, Kacprzyk J, Krawczak M, Melin P, Sotirov S, Sotirova E, Szmidt E, De Tr G, Zadrony S, editors. Novel developments in uncertainty representation and processing. Advances in intuitionistic fuzzy sets and generalized nets, vol. 401. 2016. p. 97–110.
22. Angelova N, Marinov E, Atanassov K. Intuitionistic fuzzy implications and Kolmogorovs and Lukasiewisz Tarskis axioms of logic. Notes on Intuitionistic Fuzzy Sets. 2015;21(2):35–42.
23. Atanassov K. On some intuitionistic fuzzy implications. Comptes Rendus de l'Academie bulgare des Sciences, Tome 59, 2006;(1):19–24.
24. Atanassov K. A new intuitionistic fuzzy implication from a modal type. Adv Stud Contemp Math. 2006;12(1):117–122.
25. Atanassov K. On eight new intuitionistic fuzzy implications. In: Proceedings of the 3rd International IEEE Conference on "Intelligent Systems" IS06, London, 4–6 September, 2006. p. 741–746.
26. Atanassov K. On intuitionistic fuzzy implication $\rightarrow^{\varepsilon}$ and intuitionistic fuzzy negation $\neg^{\varepsilon,\eta}$. Issues Intuit Fuzzy Sets Gen Nets. 2008;6:6–19.
27. Atanassov K. Intuitionistic fuzzy implication $\rightarrow^{\varepsilon,\eta}$ and intuitionistic fuzzy negation $\neg^{\varepsilon,\eta}$. Develop Fuzzy Sets, Intuit Fuzzy Sets, Gen Nets Relat Topics. 2008;1:1–10.
28. Atanassov K. On a new intuitionistic fuzzy implication. In: 9th Conference of the European Society for Fuzzy Logic and Technology (EUSFLAT), 30.06–03.07, Gijon, Spain, 2015. p. 1592–1597.
29. Atanassov K. On intuitionistic fuzzy implications. Issues Intuit Fuzzy Sets Gen Nets. 2016;12:1–19.
30. Atanassov K, Dimitrov D. Intuitionistic fuzzy implications and axioms for implications. Notes Intuit Fuzzy Sets. 2010;16(1):10–20.
31. Atanassov K, Kolev B. On an intuitionistic fuzzy implication from a probabilistic type. Adv Stud Contemp Math. 2006;12(1):111–116.
32. Atanassov K, Szmidt E. Remark on intuitionistic fuzzy implication $\rightarrow^{\varepsilon,\eta}$. Issues Intuit Fuzzy Sets Gen Nets. 2014;11:9–14.
33. Atanassov K, Szmidt E, Kacprzyk J. On Fodor's type of intuitionistic fuzzy implication and negation. Notes on Intuitionistic Fuzzy Sets. 2015;21(2):25–34.
34. Szmidt E, Kacprzyk J, Atanassov K. Properties of Fodor's intuitionistic fuzzy implication and negation. Notes on Intuitionistic Fuzzy Sets. 2015;21(4):6–12.
35. Szmidt E, Kacprzyk J, Atanassov K. Modal forms of Fodor's type of intuitionistic fuzzy implication. Notes on Intuitionistic Fuzzy Sets. 2015;21(5):1–5.
36. Atanassov K, Trifonov T. On a new intuitionistic fuzzy implication of Godel's type. Proc Jangjeon Math Soc. 2005;8(2):147–152.
37. Atanassov K, Trifonov T. Two new intuitionistic fuzzy implications. Adv Stud Contemp Math. 2006;13(1):69–74.
38. Atanassova L. On an intuitionistic fuzzy implication from Kleene-Dienes type. Proc Jangjeon Math Soc. 2008;11(1):69–74.
39. Atanassova L. Modifications of an intuitionistic fuzzy implication from Kleene-Dienes type. Adv Stud Contemp Math. 2008;16(2):155–160.

40. Atanassova L. New modifications of an intuitionistic fuzzy implication from Kleene-Dienes type. Part 2. Annual of Section "Informatics", vol. 1, 2008. p. 59–64.
41. Atanassova L. New modifications of an intuitionistic fuzzy implication from Kleene-Dienes type. Part 3. Adv Stud Contemp Math. 2009;18(1):33–40.
42. Atanassova L. A new intuitionistic fuzzy implication. Cybern Inf Technol. 2009;9(2):21–25.
43. Atanassova L. On some properties of intuitionistic fuzzy negation $\neg_@$. Notes on Intuitionistic Fuzzy Sets. 2009;15(1):32–35.
44. Atanassova L. On two modifications of the intuitionistic fuzzy implication $\rightarrow_@$. Notes on Intuitionistic Fuzzy Sets. 2012;18(2):26–30.
45. Atanassova L. On the modal form of the intuitionistic fuzzy implications $\rightarrow'_@$ and $\rightarrow''_@$. Issues Intuit Fuzzy Sets Gen Nets. 2013;10:5–11.
46. Atanassova L. On the intuitionistic fuzzy form of the classical implication $(A \rightarrow B) \vee (B \rightarrow A)$. Notes on Intuitionistic Fuzzy Sets. 2013;19(4):15–18.
47. Atanassova L. Remark on the intuitionistic fuzzy forms of two classical logic axioms. Part 1. Annual of Section "Informatics". 2014;7:24–27.
48. Atanassova L. Remark on the intuitionistic fuzzy forms of two classical logic axioms. Part 2. Notes Intuit Fuzzy Sets. 2014;20(4):10–13.
49. Atanassova L. Remark on Dworniczak's intuitionistic fuzzy implications. Part 1. Notes on Intuitionistic Fuzzy Sets. 2015;21(3):18–23.
50. Atanassova L. Remark on Dworniczak's intuitionistic fuzzy implications. Part 2, Annual of "Informatics" Section Union of Scientists in Bulgaria, Vol. 8, 2015 (in press).
51. Atanassova L. Remark on Dworniczak's intuitionistic fuzzy implications. Part 3. Issues Intuit Fuzzy Sets Gen Nets. 2016;12 (in press).
52. Dworniczak P. Some remarks about the L. Atanassova's paper "A new intuitionistic fuzzy implication". Cybern Inf Technol. 2010;10(3):3–9.
53. Dworniczak P. On one class of intuitionistic fuzzy implications. Cybern Inf Technol. 2010;10(4):13–21.
54. Dworniczak P. On some two-parametric intuitionistic fuzzy implication. Notes on Intuitionistic Fuzzy Sets. 2011;17(2):8–16.
55. Riecan B, Atanassov K. On a new intuitionistic fuzzy implication of Gaines-Rescher's type. Notes on Intuitionistic Fuzzy Sets. 2007;13(4):1–4.
56. Klir G, Yuan B. Fuzzy sets and fuzzy logic. Upper Saddle River: Prentice Hall; 1995.
57. Atanassov K. On some intuitionistic fuzzy negations. In: Proceedings of the First International Workshop on IFSs, Banska Bystrica, 22 September, 2005. Notes on Intuitionistic Fuzzy Sets. 2005;11(6):13–20.
58. Atanassov K. Second Zadehs intuitionistic fuzzy implication. Notes on Intuitionistic Fuzzy Sets. 2011;17(3):11–14.
59. Atanassov K. On intuitionistic fuzzy negations and Law of Excluded Middle. In: Proceedings of the 2010 IEEE International Conference on Intelligent Systems (IS 2010), London, 7–9 July 2010. p. 266–269.
60. Atanassov K, Angelova N. Maximal and minimal intuitionistic fuzzy negations. In: Atanassov K, Baczynski M, Drewniak J, Kacprzyk J, Krawczak M, Szmidt E, Wygralak M, Zadrozny S, editors. Modern approaches in fuzzy sets, intuitionistic fuzzy sets, generalized nets and related topics, Volume 1: Foundations. Warsaw: SRI-PAS; 2014. p. 51–61.
61. Atanassov K, Angelova N. On intuitionistic fuzzy negations, law for excluded middle and De Morgan's laws. Issues Intuit Fuzzy Sets Gen Nets. vol. 12, 2016 (in press).
62. Atanassov K, Dimitrov D. On the negations over intuitionistic fuzzy sets. Part 1, Annual of "Informatics" Section Union of Scientists in Bulgaria, vol. 1, 2008, p. 49–58.
63. Atanassov K, Angelova N. Properties of intuitionistic fuzzy implications and negations. In: Proceedings of the 20th International Conference on Intuitionistic Fuzzy Sets, Sofia, 2–3 Sept 2016, Notes on Intuitionistic Fuzzy Sets 2016;22(3):25–33.
64. Plisko V. A survey of propositional realizability logic. Bull Symb Logic. 2009;15(1):1–42.
65. Rose GF. Propositional calculus and realizability. Trans Am Math Soc. 1953;75:1–19.
66. Trillas E. On the law in fuzzy logic. IEEE Trans Fuzzy Syst. 2002;10(1):84–88.

67. Rasiova H, Sikorski R. The mathematics of metamathematics. Pol Acad of Sci Warszawa, 1963.
68. Tabakov M. Logics and axiomatics. Sofia: Nauka i Izkustvo; 1986 (in Bulgarian).
69. Hoormann CFA Jr. On Hauber's statement of his theorem. Notre Dame Journal of Formal Logic. 1971;12(1):86–88.
70. Sobocinski B. Lattice-theoretical and mereological forms of Hauber's law. Notre Dame Journal of Formal Logic. 1971;12(1):81–85.
71. Atanassov K. The Hauber's law is an intuitionistic fuzzy tautology. Notes on Intuitionistic Fuzzy Sets. 1997;3(2):82–84.
72. Schalkoff R. Artificial intelligence. New York: McGraw-Hill; 1990.
73. Gargov G, Atanassov K. An intuitionistic fuzzy interpretation of the basic axiom of the resolution. Notes on Intuitionistic Fuzzy Sets. 1996;2(3):20–21.
74. Nute D. Defeasible reasoning and decision support systems. Decis Support Syst. 1988;4: 97–110.
75. Atanassov K. Intuitionistic fuzzy interpretation of the conditional logic VW. In: Lakov D, editor. Proceedings of the Second Workshop on Fuzzy Based Expert Systems FUBEST'96, Sofia, 9–11 Oct 1996. p. 81–86.
76. Atanassov K. On intuitionistic fuzzy negations and De Morgan Laws. In: Proceedings of the Eleventh International Conference IPMU, Paris, 2–7 July 2006. PP. 2399–2404.

Chapter 2
Intuitionistic Fuzzy Predicate Logic

2.1 Short Remarks on Intuitionistic Fuzzy Predicate Logic

The idea for evaluation of the propositions was extended for predicates (see [1–6])
as follows (see, e.g., [7–10]).

Let x be a variable, obtaining values in set E and let $P(x)$ be a predicate with a
variable x. Let

$$V(P(x)) = \langle \mu(P(x)), \nu(P(x)) \rangle.$$

The IF-interpretations of the (intuitionistic fuzzy) quantifiers *for all* ($\forall$) and *there
exists* ($\exists$) are introduced in [7, 9, 10] by

$$V(\exists x\, P(x)) = \langle \sup_{y \in E} \mu(P(y)), \inf_{y \in E} \nu(P(y)) \rangle, \tag{2.1.1}$$

$$V(\forall x\, P(x)) = \langle \inf_{y \in E} \mu(P(y)), \sup_{y \in E} \nu(P(y)) \rangle. \tag{2.1.2}$$

If E is a finite set, then we can use the denotations

$$V(\exists x\, P(x)) = \langle \max_{y \in E} \mu(P(y)), \min_{y \in E} \nu(P(y)) \rangle, \tag{2.1.3}$$

$$V(\forall x\, P(x)) = \langle \min_{y \in E} \mu(P(y)), \max_{y \in E} \nu(P(y)) \rangle. \tag{2.1.4}$$

In general, below, we use the first forms of both quantifiers.

Their geometrical interpretations are illustrated in Figs. 2.1 and 2.2, respectively,
where $x_1, \ldots, x_5$ are possible values of variable x and $V(x_1), \ldots, V(x_5)$ are their
IF-evaluations.

The most important property of the two quantifiers is that each of them juxtaposes
to predicate P a point (exactly one per quantifier) in the IF-interpretational triangle.

© Springer International Publishing AG 2017
K.T. Atanassov, *Intuitionistic Fuzzy Logics*, Studies in Fuzziness
and Soft Computing 351, DOI 10.1007/978-3-319-48953-7_2

Fig. 2.1 Second geometrical interpretation of quantifier ∀

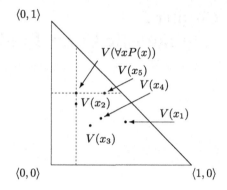

Fig. 2.2 Second geometrical interpretation of quantifier ∃

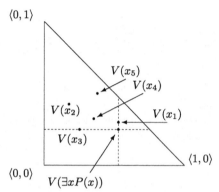

In [9, 10], for implication $\rightarrow_4$ the following two theorems are proved, where we used $\rightarrow$ instead of $\rightarrow_4$.

Theorem 2.1.1 *The logical axioms of the $\mathcal{K}$-theory (see [5]):*

(a) $A \rightarrow (B \rightarrow A)$,
(b) $(A \rightarrow (B \rightarrow C)) \rightarrow ((A \rightarrow B) \rightarrow (A \rightarrow C))$,
(c) $(\neg A \rightarrow \neg B) \rightarrow ((\neg A \rightarrow B) \rightarrow A)$,
(d) $\forall x\, A(x) \rightarrow A(t)$, *for the fixed variable* t,
(e) $\forall x (A \rightarrow B) \rightarrow (A \rightarrow \forall x B)$
     are IFTs.

Proof Assertions (a)–(c) coincide with those in Theorem 1.5.17 (IL2), (IL5) and Theorem 1.5.26, respectively. We will here prove only assertions (d) and (e).
  (d) Let the variable t be fixed.
  Then,

$$V(\forall x\, A(x) \rightarrow A(t))$$

$$= \langle \inf_x \mu(A(x)), \sup_x \nu(A(x)) \rangle \rightarrow \langle \mu(A(t)), \nu(A(t)) \rangle$$

$$= \langle \max(\sup_x \nu(A(x)), \mu(A(t))), \min(\inf_x \mu(A(x)), \nu(A(t))) \rangle$$

and

$$\max(\sup_x \nu(A(x)), \mu(A(t))) - \min(\inf_x \mu(A(x)), \nu(A(t)))$$

$$\geq \mu(A(t)) - \inf_x \mu(A(x)) \geq 0,$$

i.e., (d) is an IFT.

For (e) we sequentially obtain:

$$V(\forall x(A \to B) \to (A \to \forall x B)) = V(\forall x(A \to B)) \to V(A \to \forall x B)$$

$$= \langle \inf_x \max(\mu(B), \nu(A)), \sup_x \min(\mu(A), \nu(B)) \rangle$$

$$\to \langle \max(\nu(A), \inf_x \mu(B)), \min(\mu(A), \sup_x \nu(B)) \rangle$$

$$= \langle \max(\nu(A), \inf_x \mu(B), \sup_x \min(\mu(A), \nu(B))),$$

$$\min(\mu(A), \sup_x \nu(B), \inf_x \max(\mu(B), \nu(A))) \rangle$$

and

$$\max(\nu(A), \inf_x \mu(B), \sup_x \min(\mu(A), \nu(B))) \geq \max(\nu(A), \inf_x \mu(B))$$

$$= \inf_x \max(\mu(B), \nu(A)) \geq \min(\mu(A), \sup_x \nu(B), \inf_x \max(\mu(B), \nu(A))),$$

i.e., (e) also is an IFT.   □

Below, we list some assertions, which are theorems of the classical first order logic (see, e.g. [5]).

Theorem 2.1.2 *The following formulas are IFTs:*

(a) $(\forall x A(x) \to B) \equiv \exists x(A(x) \to B),$
(b) $\exists x A(x) \to B \equiv \forall x(A(x) \to B),$
(c) $B \to \forall x A(x) \equiv \forall x(B \to A(x)),$
(d) $B \to \exists x A(x) \equiv \exists x(B \to A(x)),$
(e) $(\forall x A \wedge \forall x B) \equiv \forall x(A \wedge B),$
(f) $(\forall x A \vee \forall x B) \to \forall x(A \vee B),$
(g) $\neg \forall x A \equiv \exists x \neg A,$

(h) $\neg\exists x A \equiv \forall x \neg A$,

(i) $\forall x \forall y A \equiv \forall y \forall x A$,

(j) $\exists x \exists y A \equiv \exists y \exists x A$,

(k) $\exists x \forall y A \rightarrow \forall y \exists x A$,

(l) $\forall x (A \rightarrow B) \rightarrow (\forall x A \rightarrow \forall x B)$.

Proof We shall use Lemma 1.5.1.

(a)

$$V(\forall x A(x) \rightarrow B)$$

$$= \langle \max(\sup_{x} \nu(A(x)), \mu(B)), \min(\inf_{x} \mu(A(x)), \nu(B)) \rangle$$

$$= \langle \sup_{x}(\max(\nu(A(x)), \mu(B))), \inf_{x}(\min(\mu(A(x)), \nu(B))) \rangle$$

$$= V(\exists x (A(x) \rightarrow B));$$

(b)

$$V(\exists x A(x) \rightarrow B)$$

$$= \langle \max(\inf_{x} \nu(A(x)), \mu(B)), \min(\sup_{x} \mu(A(x)), \nu(B)) \rangle$$

$$= \langle \inf_{x}(\max(\nu(A(x)), \mu(B))), \sup_{x}(\min(\mu(A(x)), \nu(B))) \rangle$$

$$= V(\forall x (A(x) \rightarrow B));$$

(c)

$$V(B \rightarrow \forall x A(x))$$

$$= \langle \max(\inf_{x} \mu(A(x)), \nu(B)), \min(\sup_{x} \nu(A(x)), \mu(B)) \rangle$$

$$= \langle \inf_{x}(\max(\mu(A(x)), \nu(B))), \sup_{x}(\min(\nu(A(x)), \mu(B))) \rangle$$

$$= V(\forall x (B \rightarrow A(x)));$$

(d) is proved analogically;

(e)

$$V(\forall x A \wedge \forall x B)$$

$$= \langle \min(\inf_{x} \mu(A), \min \mu(B)), \max(\sup_{x} \nu(A), \max \nu(B)) \rangle$$

$$= \langle \inf_{x}(\min(\mu(A), \mu(B))), \sup_{x}(\max(\nu(A), \nu(B))) \rangle$$

$$= V(\forall x(A \wedge B));$$

(f) is proved analogically;

(g)

$$V(\neg \forall x A) = \langle \sup_x \nu(A), \inf_x \mu(A) \rangle = V(\exists x \neg A);$$

(h) is proved analogically;

(i)

$$V(\forall x \forall y A)$$

$$= \langle \inf_x \inf_y \mu(A), \sup_x \sup_y \nu(A) \rangle$$

$$= \langle \inf_y \inf_x \mu(A), \sup_y \sup_x \nu(A) \rangle$$

$$= V(\forall y \forall x A);$$

(j) is proved analogically;

(k)

$$V(\exists x \forall y A \to \forall y \exists x A)$$

$$= \langle \sup_x \inf_y \mu(A), \inf_x \sup_y \nu(A) \rangle \to \langle \inf_y \sup_x \mu(A), \sup_y \inf_x \nu(A) \rangle$$

$$= \langle \max(\inf_y \sup_x \nu(A), \inf_x \sup_y \mu(A)), \min(\sup_x \inf_y \mu(A), \sup_y \inf_x \nu(A)) \rangle$$

and

$$\max(\inf_y \sup_x \nu(A), \inf_x \sup_y \mu(A)) - \min(\sup_x \inf_y \mu(A), \sup_y \inf_x \nu(A))$$

$$\geq \sup_y \inf_x \nu(A) - \sup_y \inf_x \nu(A) = 0,$$

i.e., $\exists x \forall y A \to \forall y \exists x A$ is an IFT;

(l) is proved analogically. $\qquad \qquad \Box$

Theorem 2.1.3 *For a predicate P and for negation $\neg_i$, $\forall x P(x) \vee \exists x \neg_i P(x)$ is an IFT for $i = 1, 3, 4, 8, 9, 11, 12, 14, 15, 18, \ldots, 23, 25, \ldots, 32, 45, 52, 53$.*

Proof Let for the variable x,

$$V(\forall x P(x)) = \langle M, n \rangle,$$

$$V(\exists x P(x)) = \langle m, N \rangle,$$

where the pairs $\langle M, n \rangle$ and $\langle m, N \rangle$ are given by either (2.1.1) and (2.1.2), or (2.1.3) and (2.1.4). Below, we will discuss the proof for three of the negations: $\neg_1$, $\neg_{12}$ and $\neg_{52}$.

For $\neg_1$ we obtain:

$$\forall x\, P(x) \vee \exists x \neg_1 P(x) = \langle m, N \rangle \vee V(\exists x \langle \nu(P(x)), \mu(P(x)) \rangle)$$

$$= \langle m, N \rangle \vee \langle N, m \rangle = \langle \max(m, N), \min(m, N) \rangle,$$

that is an IFT, because $\max(m, N) - \min(m, N) \geq 0$.

For $\neg_{12}$ we obtain:

$$\forall x\, P(x) \vee \exists x \neg_{12} P(x)$$

$$= \langle m, N \rangle \vee \langle \sup_x(\nu(P(x))(\mu(P(x)) + \nu(P(x)))),$$

$$\inf_x(\mu(P(x))(\mu(P(x)) + \mu(P(x))\nu(P(x)) + \nu(P(x))^2)) \rangle$$

$$= \langle \max(m, \sup_x(\nu(P(x))(\mu(P(x)) + \nu(P(x))))),$$

$$\min(N, \inf_x(\mu(P(x))(\mu(P(x)) + \mu(P(x))\nu(P(x)) + \nu(P(x))^2))) \rangle.$$

Then,

$$\max(m, \sup_x(\nu(P(x))(\mu(P(x)) + \nu(P(x)))))$$

$$- \min(N, \inf_x(\mu(P(x))(\mu(P(x)) + \mu(P(x))\nu(P(x)) + \nu(P(x))^2)))$$

$$\geq m - \inf_x(\mu(P(x))(\mu(P(x)) + \mu(P(x))\nu(P(x)) + \nu(P(x))^2)) \geq 0,$$

because for every two numbers $a, b \in [0, 1]$, such that $a + b \leq 1$: $a + ab + b^2 = a + b(a + b) \leq a + b \leq 1$, i.e., the expression is an IFT.

For $\neg_{52}$ we obtain:

$$\forall x\, P(x) \vee \exists x \neg_{52} P(x)$$

$$= \langle m, N \rangle \vee \langle \sup_x(\mathrm{sg}(\nu(P(x))) + \mathrm{sg}(\mu(P(x))\nu(P(x)))), m \rangle$$

$$= \langle \max(m, \sup_x(\mathrm{sg}(\nu(P(x))) + \mathrm{sg}(\mu(P(x))\nu(P(x))))), \min(N, m) \rangle,$$

which obviously is an IFT. The $(\sup - \inf)$-case is analogous.

All other checks are similar. $\square$

The link between the interpretations of quantifiers and the topological operators C (closure) and I (interior) defined over IFSs see [7] is obvious.

Open Problem 9. The basic problem which remains unsolved is related to the characterization of predicate IFTs by means of a calculus.

Following [9, 10], we mention that a partial solution of the problem of giving a calculus which generates all predicate IFTs is presented in the next theorem.

Theorem 2.1.4 *A prenex normal form A is an IFT if and only if it is a classical predicate tautology and its quantifier free matrix is a propositional IFT.*

Here, a prenex form means (see [9, 10]) a predicate formula in which all quantifiers are moved to the left. The proof is based on the fact that all predicate transformations leading to prenex forms in the classical logic are valid for the intuitionistic fuzzy case, too.

2.2 Extended Intuitionistic Fuzzy Quantifiers

In [8], we introduced the following six quantifiers and studied some of their properties.

$$V(\forall_\mu x P(x)) = \{\langle x, \inf_{y \in E} \mu(P(y)), \nu(P(x))\rangle | x \in E\},$$

$$V(\forall_\nu x P(x)) = \{\langle x, \min(1 - \sup_{y \in E} \nu(P(y)), \mu(P(x))), \sup_{y \in E} \nu(P(y))\rangle | x \in E\},$$

$$V(\exists_\mu x P(x)) = \{\langle x, \sup_{y \in E} \mu(P(y)), \min(1 - \sup_{y \in E} \mu(P(y)), \nu(P(x)))\rangle | x \in E\},$$

$$V(\exists_\nu x P(x)) = \{\langle x, \mu(P(x)), \inf_{y \in E} \nu(P(y))\rangle | x \in E\},$$

$$V(\forall_\nu^* x P(x))$$

$$= \{\langle x, \min(1 - \sup_{y \in E} \nu(P(y)), \mu(P(x))), \min(\sup_{y \in E} \nu(P(y)), 1 - \mu(P(x)))\rangle | x \in E\},$$

$$V(\exists_\mu^* x P(x))$$

$$= \{\langle x, \min(\sup_{y \in E} \mu(P(y)), 1 - \nu(P(x))), \min(1 - \sup_{y \in E} \mu(P(y)), \nu(P(x)))\rangle | x \in E\}.$$

⟨0, 1⟩
V(a)
V(c)
V(b)
⟨0, 0⟩
⟨1, 0⟩

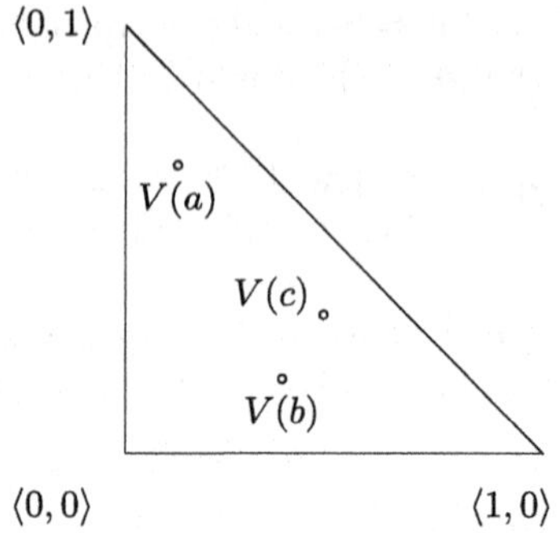

Fig. 2.3 Example of a second geometrical interpretation

⟨0, 1⟩
V(a)
V(∃μxP(x))
V(c)
V(b)
⟨0, 0⟩
⟨1, 0⟩

Fig. 2.4 Second geometrical interpretation of quantifier $\exists_\mu$

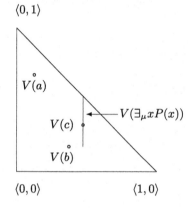

⟨0, 1⟩
V(a)
V(∃νxP(x))
V(c)
V(b)
⟨0, 0⟩
⟨1, 0⟩

Fig. 2.5 Second geometrical interpretation of quantifier $\exists_\nu$

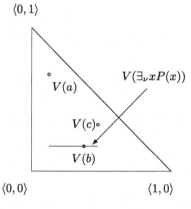

Let the possible values of the variable x be a, b, c and let their IF-evaluations $V(a)$, $V(b)$, $V(c)$ be shown on Fig. 2.3. The geometrical interpretations of the new quantifiers are shown in Figs. 2.4, 2.5, 2.6, 2.7, 2.8 and 2.9.

Fig. 2.6 Second geometrical interpretation of quantifier $\forall_\mu$

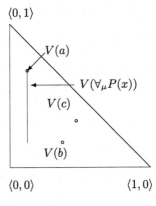

Fig. 2.7 Second geometrical interpretation of quantifier $\forall_\nu$

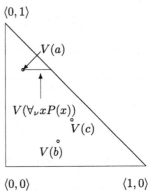

Fig. 2.8 Second geometrical interpretation of quantifier $\exists_\mu^*$

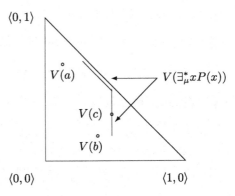

Fig. 2.9 Second geometrical
interpretation of quantifier $\forall^*_\nu$

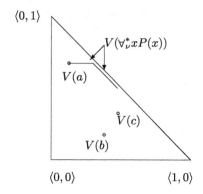

Now, we see that we can change the forms of the first two quantifiers to the forms

$$V(\forall x\, P(x)) = \{\langle x,\, \inf_{y\in E} \mu(P(y)),\, \sup_{y\in E} \nu(P(y))\rangle | x \in E\},$$

$$V(\exists x\, P(x)) = \{\langle x,\, \sup_{y\in E} \mu(P(y)),\, \inf_{y\in E} \nu(P(y))\rangle | x \in E\}.$$

Obviously, for every predicate P,

$$V(\forall x\, P(x)) \subseteq V(\forall_\mu x\, P(x)) \subseteq V(\forall_\nu x\, P(x)) \subseteq V(\exists_\nu x\, P(x))$$

$$\subseteq V(\exists_\mu x\, P(x)) \subseteq V(\exists x\, P(x))$$

and

$$V(\forall x\, P(x)) \subseteq V(\forall_\nu x\, P(x)) \subseteq V(\forall^*_\nu x\, P(x))$$

$$\subseteq V(\exists^*_\mu x\, P(x)) \subseteq V(\exists_\mu x\, P(x)) \subseteq V(\exists x\, P(x)).$$

Open Problem 10. Which implications satisfy Theorem 2.1.1(d) and (e) in Sect. 2.1?

Now, we can modify the new six operators, so as to change their set form to the
form of the first two operators.
Let a be one of the possible values for variable x. Then,

$$V((\forall_\mu x\, P(x)), a) = \langle \inf_{y\in E} \mu(P(y)), \nu(P(a))\rangle,$$

$$V((\forall_\nu x\, P(x)), a) = \langle \min(1 - \sup_{y\in E} \nu(P(y)), \mu(P(a))), \sup_{y\in E} \nu(P(y))\rangle,$$

$$V((\exists_\mu x\, P(x)), a) = \langle \sup_{y\in E} \mu(P(y)), \min(1 - \sup_{y\in E} \mu(P(y)), \nu(P(a)))\rangle,$$

$$V((\exists_\nu x P(x)), a) = \langle \mu(P(a)), \inf_{y \in E} \nu(P(y)) \rangle,$$

$$V((\forall_\nu^* x P(x)), a)$$

$$= \langle \min(1 - \sup_{y \in E} \nu(P(y)), \mu(P(a))), \min(\sup_{y \in E} \nu(P(y)), 1 - \mu(P(a))) \rangle,$$

$$V((\exists_\mu^* x P(x)), a)$$

$$= \langle \min(\sup_{y \in E} \mu(P(y)), 1 - \nu(P(a))), \min(1 - \sup_{y \in E} \mu(P(y)), \nu(P(a))) \rangle.$$

We finish this section with an example.

Let the universe comprise the members of the European Union and let for each country the degree of government approval and disapproval be known. Let the predicate $P(x)$ be "The government of country x is widely approved by the people of country x". The first quantifier $\forall$ will give the minimal degree of approval which exists in the countries of the EU, and the maximal degree of disapproval in the countries (not necessarily the same). Conversely, the second operator $\exists$ will give us the maximal degree of approval in one of these countries and the minimal degree of disapproval.

Let us assume that for some reason we do not have complete information about either the approval or disapproval for a fixed country a from the EU (but we have such information about the rest). If we are missing information about the degree of approval for a, then, the third operator $\forall_\mu$ will give us a lower bound for this degree of approval for a. The fifth operator $\exists_\mu$ will give us an upper bound for the degree of approval for a.

Conversely, if we are missing information about the degree of disapproval, the fourth operator will give us $\forall_\nu$ will give us the upper bound and the sixth $\exists_\nu$ will give us the lower bound for the degree of disapproval for a.

The seventh and eighth operators act exactly like the fourth and the fifth operators, respectively, but provide a more precise evaluation for the respective degree.

2.3   Ideas for New Types of Quantifiers

It is well known from the classical logic that for each predicate P with argument x having a finite number of interpretations $a_1, a_2, \ldots, a_n$:

$$V(\forall x P(x)) = V(P(a_1) \wedge P(a_2) \wedge \cdots \wedge P(a_n)),$$

$$V(\exists x P(x)) = V(P(a_1) \vee P(a_2) \vee \cdots \vee P(a_n)).$$

Now, following [11] and having in mind the ideas from Sect. 1.7, we can construct a lot of new quantifiers. For each new pair of conjunction and disjunction, we obtain a pair of quantifiers that have the forms

$$V(\forall_{i,j} P(x)) = V(P(a_1) \wedge_{i,j} P(a_2) \wedge_{i,j} \cdots \wedge_{i,j} P(a_n)),$$

$$V(\exists_{i,j} P(x)) = V(P(a_1) \vee_{i,j} P(a_2) \vee_{i,j} \cdots \vee_{i,j} P(a_n)),$$

where i ($1 \le i \le 185$) and j ($1 \le j \le 3$) are the indices of the respective pair of conjunction and disjunction that generates the new pair of quantifiers.

Obviously, $\forall_{4,1}$ coincides with the standard quantifier $\forall$, and $\exists_{4,1}$ coincides with the standard quantifier $\exists$.

One special case is the following: using implication $\rightarrow_{139}$ and negation $\neg_1$ we obtain for $a, b, c, d \in [0, 1]$ and $a + b, c + d \le 1$:

$$V(\langle a, b \rangle \vee_{139,3} \langle c, d \rangle) = \left\langle \frac{a+c}{2}, \frac{b+d}{2} \right\rangle = \langle a, b \rangle \wedge_{139,3} \langle c, d \rangle.$$

Therefore, if for each i: $V(P(x_i)) = \langle a_i, b_i \rangle$, then,

$$V(\forall_{139,3} x P(x)) = \left\langle \frac{\sum\limits_{i=1}^{n} a_i}{n}, \frac{\sum\limits_{i=1}^{n} b_i}{n} \right\rangle = V(\exists_{139,3} P(x)).$$

In this case, we check directly, that

$$\neg_1 \forall_{139,3} x \neg_1 P(x) = \forall_{139,3} x P(x).$$

Hence, there exists a quantifier's interpretation for which both quantifiers "$\forall$" and "$\exists$" coincide.

It is very interesting that the topological weight-center operator W (see, e.g. [12]) is an exact analogue of quantifier $\forall_{139,3}$. So, we can denote it as $W x P(x))$.

The so defined quantifiers give us the possibility to clasify all of them in two groups.

- Global quantifiers: $\forall$, $\exists$, W,
- Local quantifiers: $\forall_\mu$, $\forall_\mu$, $\forall_\nu^*$, $\exists_\mu$, $\exists_\nu$, $\exists_\mu^*$, U.

Open Problem 11. Study in details the behaviour of these quantifiers.

References

1. Barwise J, editor. Handbook of mathematical logic., Studies in Logic and the Foundations of Mathematics, Amsterdam: North Holland; 1989.
2. Crossley JN, Ash CJ, Brickhill CJ, Stillwell JC, Williams NH. What is mathematical logic?. London: Oxford University Press; 1972.

3. van Dalen D. Logic and structure. Berlin: Springer; 2013.
4. Ebbinghaus H-D, Flum J, Thomas W. Mathematical logic. 2nd ed. New York: Springer; 1994.
5. Mendelson E. Introduction to mathematical logic. Princeton: D. Van Nostrand; 1964.
6. Shoenfield JR. Mathematical logic. 2nd ed. Natick: A.K. Peters; 2001.
7. Atanassov K. Intuitionistic fuzzy sets. Heidelberg: Springer; 1999.
8. Atanassov K. On intuitionistic fuzzy logics: Results and problems. In: Atanassov K, Baczynski M, Drewniak J, Kacprzyk J, Krawczak M, Szmidt E, Wygralak M, Zadrozny S, editors. Modern approaches in fuzzy sets, intuitionistic fuzzy sets, generalized nets and related topics, Volume 1: Foundations, SRI-PAS, Warsaw, 2014. P. 23–49.
9. Atanassov K, Gargov G. Elements of intuitionistic fuzzy logic I. Fuzzy sets Syst. 1998;95(1):39–52.
10. Gargov G, Atanassov K. Two results in intuitionistic fuzzy logic. Comptes Rendus de l'Academie bulgare des Sciences Tome. 1992;45(12):29–31.
11. Atanassov, K. On intuitionistic fuzzy quantifiers. Notes Intuitionistic Fuzzy Sets. 2016;22(2): 1–12.
12. Atanassov K. On intuitionistic fuzzy sets theory. Berlin: Springer; 2012.

Chapter 3
Intuitionistic Fuzzy Modal Logics

The first step of the development of the idea of intuitionistic fuzziness (see [1]), was related to introducing an intuitionistic fuzzy interpretation of the classical (standard) modal operators *"necessity"* and *"possibility"* (see, e.g., [2–5]). In the period 1988–1993, we defined eight new operators, extending the first two ones. In the end of last and in the beginning of this century, a lot of new operators were introduced. Here, we discuss the most interesting ones of them and study their basic properties.

3.1 Intuitionistic Fuzzy Classical Modal Operators

For the formula A for which $V(A) = \langle a, b \rangle$, where $a, b \in [0, 1]$ and $a + b \leq 1$, following [1], we define the two modal operators "necessity" and "possibility":

$$V(\Box A) = \langle a, 1 - a \rangle,$$

$$V(\Diamond A) = \langle 1 - b, b \rangle,$$

respectively.

It is suitable to define the evaluation function V so that:

$$V(\Box A) = \Box V(A),$$

$$V(\Diamond A) = \Diamond V(A).$$

Two different geometrical interpretations of both operators are given in Figs. 3.1, 3.2, 3.3 and 3.4, respectively.

© Springer International Publishing AG 2017
K.T. Atanassov, *Intuitionistic Fuzzy Logics*, Studies in Fuzziness
and Soft Computing 351, DOI 10.1007/978-3-319-48953-7_3

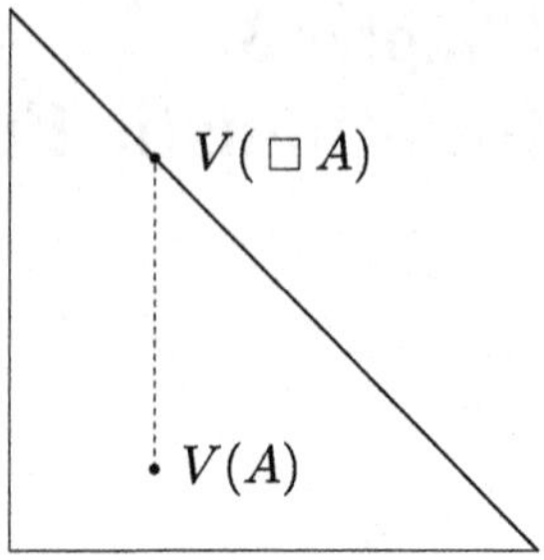

Fig. 3.1 Second geometrical
interpretation of operator □

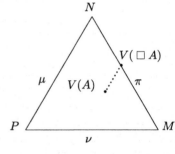

Fig. 3.2 Third geometrical
interpretation of operator □

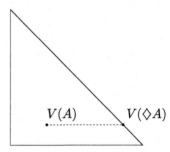

Fig. 3.3 Second geometrical
interpretation of operator ◇

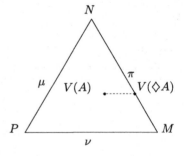

Fig. 3.4 Third geometrical
interpretation of operator ◇

It is obvious, that if p is a tautology, then, $\Box\, p$ and $\Diamond\, p$ are also tautologies. Moreover,

$$V(\Box\, p) \leq V(p) \leq V(\Diamond\, p),$$

where relation " $\leq$ " is defined as in (1.1.9).

Let everywhere below:

$$V(p) = \langle a, b \rangle, \; V(q) = \langle c, d \rangle, \; V(r) = \langle e, f \rangle,$$

where $a, b, c, d, e, f \in [0, 1], a + b \leq 1, c + d \leq 1, e + f \leq 1.$

Here, some of the most important assertions, related to the two classical (standard) modal operators, are formulated and proved for the intuitionistic fuzzy case.

First, following [1], we see that the basic properties of the (standard) modal operators in their intuitionistic fuzzy interpretations for every formula A are:

$$V(\Box\, \Box\, A) = V(\Box\, A),$$

$$V(\Box\Diamond\, A) = V(\Diamond\, A),$$

$$V(\Diamond\, \Box\, A) = V(\Box\, A),$$

$$V(\Diamond\, \Diamond\, A) = V(\Diamond\, A).$$

In classical modal logic, expressions

$$V(\neg\Box\, A) = V(\Diamond\,\neg A) \tag{3.1.1}$$

$$V(\Box\, A) = V(\neg\Diamond\,\neg A) \tag{3.1.2}$$

$$V(\neg\Diamond\, A) = V(\Box\,\neg A) \tag{3.1.3}$$

$$V(\Diamond\, A) = V(\neg\Box\,\neg A) \tag{3.1.4}$$

are (in some sense) equivalent. In the intuitionistic fuzzy case, similarly with De Morgan's Laws, the situation is different.

Theorem 3.1.1 *For every formula A,*

(a) *expression (3.1.1) is a tautology and an IFT for negations $\neg_1, \neg_2, \neg_6, \neg_8, \neg_{13},$* $\neg_{14}, \neg_{35}, \ldots, \neg_{38}, \neg_{40}, \neg_{42}, \ldots, \neg_{46}, \neg_{48}, \neg_{50}, \neg_{53},$
(b) *expression (3.1.2) is a tautology and an IFT for negations $\neg_1, \neg_8, \neg_{53},$*
(c) *expression (3.1.3) is a tautology and an IFT for negations $\neg_1, \neg_3, \ldots, \neg_6, \neg_{11},$* $\neg_{14}, \neg_{35}, \ldots, \neg_{37}, \neg_{39}, \neg_{41}, \ldots, \neg_{45}, \neg_{47}, \neg_{49}, \neg_{50}, \neg_{51},$
(d) *expression (3.1.4) is a tautology and an IFT for negations $\neg_1, \neg_3, \neg_4.$*

This is the first case, when a given expression is a tautology in all cases when it is an IFT. As we saw in the two previous chapters, only small number of IFTs are tautologies.

Theorem 3.1.2 *For every formula A, each of the expressions*

$$\Box A \rightarrow A,$$

$$A \rightarrow \Diamond A,$$

$$\Box A \rightarrow \Diamond A$$

is

(a) *a tautology for implications* $\rightarrow_2, \rightarrow_3, \rightarrow_5, \rightarrow_8, \rightarrow_{11}, \rightarrow_{14}, \rightarrow_{15}, \rightarrow_{20}, \rightarrow_{23}, \rightarrow_{24},$ $\rightarrow_{27}, \rightarrow_{31}, \rightarrow_{32}, \rightarrow_{34}, \rightarrow_{37}, \rightarrow_{40}, \rightarrow_{42}, \rightarrow_{47}, \rightarrow_{48}, \rightarrow_{49}, \rightarrow_{52}, \rightarrow_{55}, \rightarrow_{57},$ $\rightarrow_{62}, \rightarrow_{63}, \rightarrow_{65}, \rightarrow_{68}, \rightarrow_{69}, \rightarrow_{74}, \rightarrow_{77}, \rightarrow_{79}, \rightarrow_{83}, \rightarrow_{84}, \rightarrow_{88}, \rightarrow_{92}, \rightarrow_{93},$ $\rightarrow_{97}, \rightarrow_{176}, \ldots, \rightarrow_{185},$

(b) *an IFT for implications* $\rightarrow_1, \ldots, \rightarrow_9, \rightarrow_{11}, \ldots, \rightarrow_{15}, \rightarrow_{17}, \rightarrow_{18}, \rightarrow_{20}, \ldots,$ $\rightarrow_{24}, \rightarrow_{27}, \ldots, \rightarrow_{38}, \rightarrow_{40}, \rightarrow_{42}, \rightarrow_{44}, \ldots, \rightarrow_{53}, \rightarrow_{55}, \rightarrow_{57}, \rightarrow_{59}, \ldots, \rightarrow_{66},$ $\rightarrow_{68}, \rightarrow_{69}, \rightarrow_{71}, \rightarrow_{72}, \rightarrow_{74}, \ldots, \rightarrow_{77}, \rightarrow_{79}, \ldots, \rightarrow_{85}, \rightarrow_{88}, \ldots, \rightarrow_{94}, \rightarrow_{97},$ $\ldots, \rightarrow_{139}, \rightarrow_{141}, \rightarrow_{146}, \ldots, \rightarrow_{170}, \rightarrow_{176}, \ldots, \rightarrow_{185}.$

Theorem 3.1.3 *For every two formulas A and B, each of the equalities*

$$V(\Diamond(A \vee B)) = V(\Diamond A \vee \Diamond B) \tag{3.1.5}$$

$$V(\Box(A \wedge B)) = V(\Box A \wedge \Box B), \tag{3.1.6}$$

holds for the disjunction and conjunction, defined by (1.1.4) and (1.1.5).

Open Problem 12 Which other disjunctions and conjunctions, existing of which is discussed in Sect. 1.7, satisfy (3.1.5) and (3.1.6)?

Theorem 3.1.4 *For every two formulas A and B, each of the expressions*

$$\Box A \vee \Box B \rightarrow \Box(A \vee B)$$

$$\Diamond(A \wedge B) \rightarrow \Diamond A \wedge \Diamond B$$

is

(a) *a tautology for implications* $\rightarrow_2, \rightarrow_3, \rightarrow_5, \rightarrow_8, \rightarrow_{11}, \rightarrow_{14}, \rightarrow_{15}, \rightarrow_{20},$ $\rightarrow_{23}, \rightarrow_{24}, \rightarrow_{27}, \rightarrow_{31}, \rightarrow_{32}, \rightarrow_{34}, \rightarrow_{37}, \rightarrow_{40}, \rightarrow_{42}, \rightarrow_{47}, \ldots, \rightarrow_{49}, \rightarrow_{52},$ $\rightarrow_{55}, \rightarrow_{57}, \rightarrow_{62}, \rightarrow_{63}, \rightarrow_{65}, \rightarrow_{68}, \rightarrow_{69}, \rightarrow_{74}, \rightarrow_{77}, \rightarrow_{79}, \rightarrow_{83}, \rightarrow_{84}, \rightarrow_{88},$ $\rightarrow_{92}, \rightarrow_{93}, \rightarrow_{97}, \rightarrow_{176}, \ldots, \rightarrow_{185},$

(b) an IFT for implications $\rightarrow_1, \ldots, \rightarrow_9, \rightarrow_{11}, \ldots, \rightarrow_{15}, \rightarrow_{17}, \rightarrow_{18}, \rightarrow_{20}, \ldots,$
$\rightarrow_{24}, \rightarrow_{27}, \ldots, \rightarrow_{38}, \rightarrow_{40}, \rightarrow_{42}, \rightarrow_{44}, \ldots, \rightarrow_{53}, \rightarrow_{55}, \rightarrow_{57}, \rightarrow_{59}, \ldots, \rightarrow_{66},$
$\rightarrow_{68}, \rightarrow_{69}, \rightarrow_{71}, \rightarrow_{72}, \rightarrow_{74}, \ldots, \rightarrow_{77}, \rightarrow_{79}, \ldots, \rightarrow_{85}, \rightarrow_{88}, \ldots, \rightarrow_{94}, \rightarrow_{97},$
$\ldots, \rightarrow_{139}, \rightarrow_{141}, \rightarrow_{146}, \ldots, \rightarrow_{170}, \rightarrow_{176}, \ldots, \rightarrow_{185}.$

Theorem 3.1.5 *For every two formulas A and B, the expression*

$$\Box (A \rightarrow B) \rightarrow (\Box A \rightarrow \Box B)$$

is

(a) a tautology for implications $\rightarrow_2, \rightarrow_3, \rightarrow_5, \rightarrow_8, \rightarrow_{11}, \rightarrow_{14}, \rightarrow_{20}, \rightarrow_{24},$
$\rightarrow_{25}, \rightarrow_{27}, \rightarrow_{29}, \rightarrow_{47}, \ldots, \rightarrow_{49}, \rightarrow_{52}, \rightarrow_{55}, \rightarrow_{57}, \rightarrow_{58}, \rightarrow_{60}, \rightarrow_{69}, \rightarrow_{77},$
$\rightarrow_{79}, \rightarrow_{81}, \rightarrow_{92}, \rightarrow_{93}, \rightarrow_{97}, \rightarrow_{99}, \rightarrow_{177}, \rightarrow_{179}, \rightarrow_{181}, \rightarrow_{182}, \rightarrow_{184},$
(b) an IFT for implications $\rightarrow_1, \ldots, \rightarrow_9, \rightarrow_{11}, \ldots, \rightarrow_{14}, \rightarrow_{17}, \rightarrow_{18}, \rightarrow_{20},$
$\rightarrow_{21}, \rightarrow_{24}, \rightarrow_{25}, \rightarrow_{27}, \ldots, \rightarrow_{29}, \rightarrow_{46}, \ldots, \rightarrow_{53}, \rightarrow_{55}, \rightarrow_{57}, \ldots, \rightarrow_{61}, \rightarrow_{64},$
$\rightarrow_{66}, \rightarrow_{69}, \rightarrow_{71}, \rightarrow_{72}, \rightarrow_{75}, \ldots, \rightarrow_{77}, \rightarrow_{79}, \ldots, \rightarrow_{81}, \rightarrow_{91}, \ldots, \rightarrow_{94}, \rightarrow_{97},$
$\ldots, \rightarrow_{102}, \rightarrow_{108}, \ldots, \rightarrow_{113}, \rightarrow_{118}, \rightarrow_{120}, \ldots, \rightarrow_{128}, \rightarrow_{134}, \ldots, \rightarrow_{137}, \rightarrow_{139},$
$\rightarrow_{141}, \rightarrow_{147}, \rightarrow_{149}, \ldots, \rightarrow_{154}, \rightarrow_{156}, \rightarrow_{158}, \ldots, \rightarrow_{162}, \rightarrow_{165}, \ldots, \rightarrow_{167},$
$\rightarrow_{169}, \rightarrow_{177}, \rightarrow_{179}, \rightarrow_{181}, \rightarrow_{182}, \rightarrow_{184}.$

Proof (a) Let, for example, the implication be considered in the variant $\rightarrow_{11}$. Then,

$$V(\Box (A \rightarrow B) \rightarrow (\Box A \rightarrow \Box B))$$
$$= \Box \langle 1 - (1 - c)\mathrm{sg}(a - c), d\mathrm{sg}(a - c)\mathrm{sg}(d - b) \rangle \rightarrow (\langle a, 1 - a \rangle \rightarrow \langle c, 1 - c \rangle)$$
$$= \langle 1 - (1 - c)\mathrm{sg}(a - c), (1 - c)\mathrm{sg}(a - c) \rangle$$
$$\quad \rightarrow \langle 1 - (1 - c)\mathrm{sg}(a - c), (1 - c)\mathrm{sg}(a - c)^2 \rangle$$
$$= (1 - (1 - (1 - (1 - c)\mathrm{sg}(a - c))))\mathrm{sg}(a - c)$$
$$\quad \mathrm{sg}((1 - (1 - c)\mathrm{sg}(a - c)) - 1 - (1 - c)\mathrm{sg}(a - c)),$$
$$\quad (1 - c)\mathrm{sg}(a - c)\mathrm{sg}(((1 - c)\mathrm{sg}(a - c)) - (1 - c)\mathrm{sg}(a - c))^2 \rangle$$
$$= (1 - (1 - c)\mathrm{sg}(a - c))\mathrm{sg}(a - c)\mathrm{sg}(0), (1 - c)\mathrm{sg}(a - c)\mathrm{sg}(0) \rangle$$
$$= \langle 1, 0 \rangle.$$

This completes the proof. □

Theorem 3.1.6 *For every two formulas A and B, the expression*

$$(\Box (A \rightarrow B) \wedge \Box A) \rightarrow \Box B$$

is

(a) a tautology for implications $\rightarrow_2, \rightarrow_3, \rightarrow_8, \rightarrow_{11}, \rightarrow_{14}, \rightarrow_{15}, \rightarrow_{20}, \rightarrow_{24},$
$\rightarrow_{25}, \rightarrow_{27}, \rightarrow_{29}, \rightarrow_{47}, \rightarrow_{48}, \rightarrow_{52}, \rightarrow_{55}, \rightarrow_{57}, \rightarrow_{58}, \rightarrow_{60}, \rightarrow_{77}, \rightarrow_{79}, \rightarrow_{81},$
$\rightarrow_{92}, \rightarrow_{97}, \rightarrow_{99}, \rightarrow_{177}, \rightarrow_{179},$
(b) an IFT for implications $\rightarrow_1, \ldots, \rightarrow_9, \rightarrow_{11}, \ldots, \rightarrow_{15}, \rightarrow_{17}, \rightarrow_{18}, \rightarrow_{20},$
$\rightarrow_{21}, \rightarrow_{24}, \rightarrow_{25}, \rightarrow_{27}, \ldots, \rightarrow_{29}, \rightarrow_{46}, \ldots, \rightarrow_{53}, \rightarrow_{55}, \rightarrow_{57}, \ldots, \rightarrow_{61}, \rightarrow_{64},$

$\rightarrow_{66}, \rightarrow_{69}, \rightarrow_{71}, \ldots, \rightarrow_{73}, \rightarrow_{75}, \ldots, \rightarrow_{77}, \rightarrow_{79}, \ldots, \rightarrow_{81}, \rightarrow_{91}, \ldots, \rightarrow_{94},$
$\rightarrow_{96}, \ldots, \rightarrow_{102}, \rightarrow_{106}, \ldots, \rightarrow_{113}, \rightarrow_{118}, \ldots, \rightarrow_{128}, \rightarrow_{134}, \ldots, \rightarrow_{138},$
$\rightarrow_{151}, \rightarrow_{158}, \rightarrow_{161}, \rightarrow_{166}, \rightarrow_{167}, \rightarrow_{169}, \rightarrow_{177}, \rightarrow_{179}, \rightarrow_{181}, \rightarrow_{182}, \rightarrow_{184}.$

Theorem 3.1.7 *For every two formulas A and B, the expression*

$$\Box A \rightarrow (\Box (A \rightarrow B) \rightarrow \Box B)$$

is

(a) *a tautology for implications* $\rightarrow_3, \rightarrow_5, \rightarrow_{11}, \rightarrow_{14}, \rightarrow_{20}, \rightarrow_{25}, \rightarrow_{27}, \rightarrow_{29},$
   $\rightarrow_{48}, \rightarrow_{49}, \rightarrow_{57}, \rightarrow_{58}, \rightarrow_{60}, \rightarrow_{77}, \rightarrow_{79}, \rightarrow_{81}, \rightarrow_{97}, \rightarrow_{99}, \rightarrow_{181}, \rightarrow_{182}, \rightarrow_{184},$
(b) *an IFT for implications* $\rightarrow_1, \rightarrow_3, \ldots, \rightarrow_7, \rightarrow_9, \rightarrow_{11}, \ldots, \rightarrow_{14}, \rightarrow_{17}, \rightarrow_{18},$
   $\rightarrow_{20}, \rightarrow_{21}, \rightarrow_{25}, \rightarrow_{27}, \ldots, \rightarrow_{29}, \rightarrow_{46}, \rightarrow_{48}, \ldots, \rightarrow_{51}, \rightarrow_{53}, \rightarrow_{57}, \rightarrow_{58}, \rightarrow_{60},$
   $\rightarrow_{61}, \rightarrow_{64}, \rightarrow_{66}, \rightarrow_{71}, \rightarrow_{72}, \rightarrow_{75}, \ldots, \rightarrow_{77}, \rightarrow_{79}, \ldots, \rightarrow_{81}, \rightarrow_{91}, \ldots, \rightarrow_{94},$
   $\rightarrow_{97}, \ldots, \rightarrow_{102}, \rightarrow_{107}, \ldots, \rightarrow_{113}, \rightarrow_{118}, \rightarrow_{120}, \ldots, \rightarrow_{122}, \rightarrow_{124}, \ldots, \rightarrow_{128},$
   $\rightarrow_{134}, \ldots, \rightarrow_{137}, \rightarrow_{151}, \rightarrow_{158}, \rightarrow_{161}, \rightarrow_{166}, \rightarrow_{167}, \rightarrow_{169}, \rightarrow_{181}, \rightarrow_{182}, \rightarrow_{184}.$

An interesting open problem here is the following.

Open Problem 13 Determine for which pair of implications $(\rightarrow_i, \rightarrow_j)$ and for every two formulas A and B, the following equality is valid:

$$V(A \rightarrow_i B) = V(\Box (A \rightarrow_j B)).$$

We remind that for each evaluation function V and for each formula A such that $V(x) = \langle a, b \rangle$, A is "intuitionistic fuzzy sure" (IF-sure), iff $a \geq 0.5 \geq b$.

Let for the variable x, for which $V(x) = \langle a, b \rangle$, it is valid that

$$V(\Box x) = \langle a, 1 - a \rangle,$$

$$V(\Diamond x) = \langle 1 - b, b \rangle$$

(see Fig. 3.5).

Fig. 3.5 Second geometrical
interpretation of operators
$\Box$ and $\Diamond$

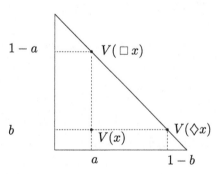

Following and extending [6], we give the following theorem.

Theorem 3.1.8 (Locality of intuitionistic fuzzy sure) *For every formula A, if $A(\Box x)$ and $A(\Diamond x)$ are IF-sure, then, for every y, for which*

$$V(\Box x) \leq V(y) \leq V(\Diamond x), \qquad (3.1.7)$$

$A(y)$ is IF-sure.

Proof Let

$$V(A(\Box x)) = \langle a(x), b(x) \rangle = \langle a, b \rangle,$$

$$V(A(\Diamond x)) = \langle c(x), d(x) \rangle = \langle c, d \rangle,$$

$$V(A(y)) = \langle \alpha(y), \beta(y) \rangle = \langle \alpha, \beta \rangle.$$

We shall prove the assertion by induction on the complexity of the formula A. Let A be a variable, i.e., $A(x) = x$. Then,

$$a(\Box x) \geq 0.5 \geq b(\Box x) = 1 - a(\Box x)$$

$$1 - d(\Diamond x) = c(\Diamond x) \geq 0.5 \geq d(\Diamond x)$$

and from (3.1.7) it follows that

$$\alpha - \beta \geq a - b \geq 0,$$

i.e., $A(y)$ is IF-sure.

Let $A = P \wedge Q$, where for P and Q the assertion is valid. Then,

$$\alpha - \beta = \min(\mu(P(y)), \mu(Q(y))) - \max(\nu(P(y)), \nu(Q(y))) \geq 0.5 - 0.5 = 0.$$

For $A = P \vee Q$, the check is similar.

Let $A(x) = \forall z P(x, z)$, where the assertion is valid for $P(x, z)$. For every y, for which (3.1.7) is valid, and for every z:

$$\mu(P(y, z)) \geq 0.5 \geq \nu(P(y, z))$$

by assumption. Then,

$$\alpha - \beta = \min \mu(P(x, z)) - \max(P(\nu(x, z))) \geq 0.5 - 0.5 = 0.$$

For $A(x) = \exists z P(x, z)$ the assertion is proved analogically.

Let $A(x) = \Box P(x)$, where the assertion is valid for $P(x)$. Then, we obtain directly, that

$$\alpha - \beta \geq a - b \geq 0.5 - 0.5 = 0.$$

When $A(x) = \Diamond P(x)$ the assertion is proved analogically.   □

Corollary 3.1.1 (Locality of intuitionistic fuzzy truth) *For every formula A, if $A(\Box x)$ and $A(\Diamond x)$ are tautologies, then, for every y, for which (3.1.7) is valid, $A(y)$ is a tautology.*

The case of IFTs is more complicated.

It can be established that if A and B are IFTs, then, $A \wedge B$ is not necessarily an IFT. For example, let $p \vee \neg p$ and $q \vee \neg q$ for different propositions p and q be IFTs, then

$$V(p \vee \neg p) = \langle \max(\mu(p), \mu(\neg p)), \min(\nu(p), \nu(\neg p)) \rangle$$
$$= \langle \max(\mu(p), \nu(p)), \min(\nu(p), \mu(\neg p)) \rangle.$$

Nevertheless, the form

$$A = (p \vee \neg p) \wedge (q \vee \neg q)$$

is not always an IFT. Take e.g.:

$$V(p) = \langle 0.4, 0.4 \rangle,$$

$$V(q) = \langle 0.2, 0.2 \rangle.$$

Then,

$$V(p \vee \neg p) = \langle 0.4, 0.4 \rangle,$$

$$V(q \vee \neg q) = \langle 0.2, 0.2 \rangle,$$

but

$$V(A) = \langle 0.2, 0.4 \rangle.$$

A Conjunctive Normal Form (CNF) A is of the sort $D_1 \wedge \cdots \wedge D_m$, where $D_i = l_{i,1} \vee \cdots \vee l_{i,k_i}$ is a clause of literals. A literal is either a propositional variable (e.g., p) or a negated variable (in the same case $- \neg p$). The literals p and $\neg p$ are called opposite. Two clauses C and D are called *connected* if they contain a common variable occurring in opposite literals (e.g., p in C and $\neg p$ in D).

Lemma 3.1.1 *A clause C is an IFT if and only if it is a classical two-valued tautology iff C contains a pair of opposite literals.*

Lemma 3.1.2 *A conjunction of two literals C and D, which are IFTs, is an IFT iff they are connected.*

Proof If C and D are IFTs and are connected, then, $C \wedge D$ is an IFT. Consider an arbitrary V. Let $C = p \vee A$, $D = \neg p \vee B$, where $V(A) = \langle a, b \rangle$, $V(B) = \langle c, d \rangle$ and let $V(p) = \langle \mu, \nu \rangle$. Then,

$$\max(\mu, a) \geq \mu \geq \min(\mu, d)$$
$$\max(\nu, c) \geq \nu \geq \min(\nu, b)$$
$$\max(\mu, a) \geq \min(\nu, b)$$
$$\max(\nu, c) \geq \min(\mu, d)$$

and from

$$V(C \wedge D) = \langle \min(\max(\mu, a), \max(\nu, c)), \max(\min(\nu, b), \min(\mu, d)) \rangle$$

and

$$\min(\max(\mu, a), \max(\nu, c)) \geq \max(\min(\nu, b), \min(\mu, d))$$

it follows that $C \wedge D$ is an IFT.

Let C and D be two IFT clauses. Let us define the following evaluation function W: For variables p which occur in both positive and negative literals in C let $W(p) = \langle 0.2, 0.2 \rangle$. For a variable q that appears in both the positive and the negative forms in D: $W(q) = \langle 0.4, 0.4 \rangle$. Note that the sets of such variables are disjoint. For variables which occur positively in C or D, let W be $\langle 0.2, 0.4 \rangle$ and for variables occurring negatively in C or $D - \langle 0.4, 0.2 \rangle$. It is a simple check which shows that $W(C \wedge D) = \langle 0.2, 0.4 \rangle$. Thus, the conjunction of C and D is not an IFT. □

A CNF A is called *totally connected* if every pair of clauses C, D in it is connected.

Lemma 3.1.3 *A CNF A is an IFT iff all clauses in it are IFTs and A is totally connected.*

Proof Assume that all clauses of A are IFTs and that A is totally connected. If we assume that for a some evaluation function W:

$$W(A) = \langle \mu, \nu \rangle$$

is such that $\mu < \nu$, then, it can be easily seen that there is a pair of clauses C and D of A such that $C \wedge D$ is already not an IFT (due to W) – but this is impossible by Lemma 3.1.2. In the opposite direction: if at least two clauses in A are not connected, then their conjunction will not be an IFT, hence, A will not be an IFT either. □

Theorem 3.1.9 (Locality of IFT) *For every connected formula A, if $A(\Box x)$ and $A(\Diamond x)$ are IFTs, then, for every y, for which (3.1.7) is valid, $A(y)$ is an IFT.*

Proof The proof is similar to the proof of Theorem 3.1.8, but in the case when formula A is a conjunction of two connected formulas, we use Lemma 3.1.2 □

Now, we discuss the basic relations between the quantifiers ∀ and ∃ and the two modal operators □ and ◇.

The equalities from [7], Sect. 1.6 can be transformed for both modal operators and both quantifiers, as follows.

Theorem 3.1.10 *Let A be a formula and x be a variable. Then,*

(a) $V(\forall x \,\square\, A) = V(\square \,\forall x\, A)$,
(b) $V(\exists x \,\square\, A) = V(\square \,\exists x\, A)$,
(c) $V(\forall x \diamond A) = V(\diamond \forall x\, A)$,
(d) $V(\exists x \diamond A) = V(\diamond \exists x\, A)$.

Proof Let us check the validity of (a):

$$V(\forall x \,\square\, A)$$
$$= \langle \min_x \mu(A), \max_x(1 - \mu(A)) \rangle$$
$$= \langle \min_x \mu(A), 1 - \min_x \mu(A) \rangle$$
$$= V(\square \,\forall x\, A).$$

Equalities (b)–(d) are proved analogically.   □

Theorem 3.1.11 *Let A be a formula and x be a variable. Then,*

(a) $V(\square \exists x \,\square\, A) = V(\diamond \,\exists x \,\square\, A) = V(\neg \,\square \,\forall x \diamond \,\neg A) = V(\neg \diamond \,\forall x \diamond \,\neg A)$,
(b) $V(\square \exists x \diamond A) = V(\diamond \,\exists x \diamond A) = V(\neg \,\square \,\forall x \,\square \,\neg A) = V(\neg \diamond \,\forall x \,\square \,\neg A)$,
(c) $V(\square \forall x \,\square\, A) = V(\diamond \,\forall x \,\square\, A) = V(\neg \,\square \,\exists x \diamond \,\neg A) = V(\neg \diamond \,\exists x \diamond \,\neg A)$,
(d) $V(\square \forall x \diamond A) = V(\diamond \,\forall x \diamond A) = V(\neg \,\square \,\exists x \,\square \,\neg A) = V(\neg \diamond \,\exists x \,\square \,\neg A)$,
(e) $V(\square \exists x \,\square \,\neg A) = V(\diamond \,\exists x \,\square \,\neg A) = V(\neg \,\square \,\forall x \diamond A) = V(\neg \diamond \,\forall x \diamond A)$,
(f) $V(\square \exists x \diamond \,\neg A) = V(\diamond \,\exists x \diamond \,\neg A) = V(\neg \,\square \,\forall x \,\square A) = V(\neg \diamond \,\forall x \,\square A)$,
(g) $V(\square \forall x \,\square \,\neg A) = V(\diamond \,\forall x \,\square \,\neg A) = V(\neg \,\square \,\exists x \diamond A) = V(\neg \diamond \,\exists x \diamond A)$,
(h) $V(\square \forall x \diamond \,\neg A) = V(\diamond \,\forall x \diamond \,\neg A) = V(\neg \,\square \,\exists x \,\square A) = V(\neg \diamond \,\exists x \,\square A)$.

Proof Let us check the validity of (a)

$$V(\square \exists x \,\square\, A)$$
$$= \square \exists x \,\square\, V(A)$$
$$= \square \exists x \langle \mu(A), 1 - \mu(A) \rangle$$
$$= \square \langle \max_x \mu(A), \min_x(1 - \mu(A)) \rangle$$
$$= \langle \max_x \mu(A), 1 - \max_x \mu(A) \rangle;$$

$$V(\diamond \exists x \,\square\, A)$$
$$= \diamond \langle \max_x(\mu(A)), \min_x(1 - \mu(A)) \rangle$$
$$= \langle 1 - \min_x(1 - \mu(A)), \min_x(1 - \mu(A)) \rangle$$
$$= \langle \max_x(\mu(A)), 1 - \max_x(\mu(A)) \rangle;$$

$$V(\neg \,\square\, \forall x \diamond \neg A) = \neg \,\square\, \forall x \diamond \langle \nu(A), \mu(A) \rangle = \neg \,\square\, \forall x \langle 1 - \mu(A), \mu(A) \rangle$$
$$= \neg \,\square\, \langle \min_x(1 - \mu(A)), \max_x(\mu(A)) \rangle$$
$$= \neg \langle \min_x(1 - \mu(A)), 1 - \min_x(1 - \mu(A)) \rangle$$
$$= \langle 1 - \min_x(1 - \mu(A)), \min_x(1 - \mu(A)) \rangle$$
$$= \langle \max_x \mu(A), 1 - \max_x \mu(A) \rangle$$
$$= V(\neg \diamond \forall x \diamond \neg A)$$
$$= \neg \diamond \langle \min_x(1 - \mu(A)), \max_x \mu(A) \rangle$$
$$= \neg \langle 1 - \max_x \mu(A), \max_x mu(A) \rangle$$
$$= \langle \max_x \mu(A), 1 - \max_x \mu(A) \rangle.$$

Equalities (b)–(h) are proved analogically. □

Let for a fixed formula A and for a variable x:

$$S(A) = \{\square \exists x \,\square\, A, \diamond \exists x \,\square\, A, \neg(\square \forall x \diamond \neg A), \neg(\diamond \forall x \diamond \neg A)\},$$

$$T(A) = \{\square \exists x \diamond A, \diamond \exists x \diamond A, \neg(\square \forall x \,\square\, \neg A), \neg(\diamond \forall x \,\square\, \neg A)\},$$

$$U(A) = \{\square \forall x \,\square\, A, \diamond \forall x \,\square\, A, \neg(\square \exists x \diamond \neg A), \neg(\diamond \exists x \diamond \neg A)\},$$

$$V(A) = \{\square \forall x \diamond A, \diamond \forall x \diamond A, \neg(\square \exists x \,\square\, \neg A), \neg(\diamond \exists x \,\square\, \neg A)\},$$

$$W(A) = \{\square \exists x \,\square\, \neg A, \diamond \exists x \,\square\, \neg A, \neg(\square \forall x \diamond A), \neg(\diamond \forall x \diamond A)\},$$

$$X(A) = \{\square \exists x \diamond \neg A, \diamond \exists x \diamond \neg A, \neg(\square \forall x \,\square\, A), \neg(\diamond \forall x \,\square\, A)\},$$

$$Y(A) = \{\square \forall x \,\square\, \neg A, \diamond \forall x \,\square\, \neg A, \neg(\square \exists x \diamond A), \neg(\diamond \exists x \diamond A)\},$$

$$Z(A) = \{\square \forall x \diamond \neg A, \diamond \forall x \diamond \neg A, \neg(\square \exists x \,\square\, A), \neg(\diamond \exists x \,\square\, A)\}.$$

Having in mind Theorem 3.1.11, we can prove the following theorem.

Theorem 3.1.12 *Let A be a formula and x be a variable. Then,*

(a) if $P \in S(A)$ and $Q \in T(A)$, then, $V(P) \leq V(\exists x A) \leq V(Q)$,
(b) if $P \in U(A)$ and $Q \in V(A)$, then, $V(P) \leq V(\forall x A) \leq V(Q)$,
(c) if $P \in W(A)$ and $Q \in X(A)$, then, $V(P) \leq V(\forall x A) \leq V(Q)$,
(d) if $P \in Y(A)$ and $Q \in Z(A)$, then, $V(P) \leq V(\exists x A) \leq V(Q)$,

where $V(X) \leq V(Y)$ for the formulas X and Y if and only if $\mu(X) \leq \mu(Y)$ and $\nu(X) \geq \nu(Y)$.

Finally, following [8], we discuss another modal operator, without an analogue in modal logic.

For the formula A, for which $V(A) = \langle a, b \rangle$, where $a, b \in [0, 1]$ and $a + b \leq 1$, we define the new modal operator "$\bigcirc$" by:

$$V(\bigcirc A) = \left\langle \frac{a}{a+b}, \frac{b}{a+b} \right\rangle.$$

Obviously, the pair $\langle \frac{a}{a+b}, \frac{b}{a+b} \rangle$ is an intuitionistic fuzzy pair and more particularly – a fuzzy pair, because

$$\frac{a}{a+b} + \frac{b}{a+b} = 1.$$

The new operator has the following more interesting properties.

Theorem 3.1.13 *For every formula A:*

(a) $\bigcirc \square A = \square A$,

(b) $\bigcirc \Diamond A = \Diamond A$,

(c) $\square \bigcirc A = \bigcirc A$,

(d) $\Diamond \bigcirc A = \bigcirc A$,

(e) $\bigcirc \bigcirc A = \bigcirc A$.

Proof (e) For formula A we obtain:

$$V(\bigcirc \bigcirc A) = \bigcirc \left\langle \frac{a}{a+b}, \frac{b}{a+b} \right\rangle = \left\langle \frac{\frac{a}{a+b}}{\frac{a}{a+b} + \frac{b}{a+b}}, \frac{\frac{b}{a+b}}{\frac{a}{a+b} + \frac{b}{a+b}} \right\rangle = V(\bigcirc A).$$

This completes the proof. The rest assertions in (a)–(d) are proved analogically.   □

Theorem 3.1.14 *For every formula A:*

(a) Only negation $\neg_1$ satisfies equality

$$\neg \bigcirc \neg A = \bigcirc A,$$

(b) Only negations $\neg_1, \neg_2, \neg_{11}, \neg_{18}, \neg_{53}$ satisfy equality

$$\neg \bigcirc A = \bigcirc \neg A.$$

Theorem 3.1.15 *For every two formulas A and B and for the disjunction and conjunction defined by (1.1.4) and (1.1.5):*

$$V(\bigcirc(A \wedge B)) \leq V(\bigcirc A \wedge \bigcirc B),$$

$$V(\bigcirc(A \vee B)) \geq V(\bigcirc A \vee \bigcirc B).$$

Proof Let A and B be two formulas, so that $V(A) = \langle a, b \rangle$, $V(B) = \langle c, d \rangle$, $a, b, c,$ $d \in [0, 1]$ and $a + b \leq 1$, $c + d \leq 1$. Then,

$$V(\bigcirc(A \wedge B)) = \bigcirc(\langle a, b \rangle \wedge \langle c, d \rangle) = \bigcirc\langle \min(a, c), \max(b, d) \rangle$$

$$= \left\langle \frac{\min(a, c)}{\min(a, c) + \max(b, d)}, \frac{\max(b, d)}{\min(a, c) + \max(b, d)} \right\rangle.$$

and

$$V(\bigcirc A \wedge \bigcirc B) = \bigcirc\langle a, b \rangle \wedge \bigcirc\langle c, d \rangle$$

$$= \left\langle \frac{a}{a+b}, \frac{b}{a+b} \right\rangle \wedge \left\langle \frac{c}{c+d}, \frac{d}{c+d} \right\rangle$$

$$= \left\langle \min\left(\frac{a}{a+b}, \frac{c}{c+d}\right), \max\left(\frac{b}{a+b}, \frac{d}{c+d}\right) \right\rangle.$$

First, we prove the validity of the following inequality. For every three real numbers $a, b, c \in [0, 1]$, if $a \geq c$, then:

$$\frac{a}{a+b} \geq \frac{c}{c+b}. \tag{3.1.8}$$

Obviously, the inequality is valid in the form of an equality, when $a = c$. Let $a > c$. Then, sequentially, we obtain:

$$\frac{a}{a+b} - \frac{c}{c+b} = \frac{ab - bc}{(a+b)(c+b)} = \frac{b(a-c)}{(a+b)(c+b)} > 0,$$

i.e., (3.1.6) is valid.

Now, we check the validity of inequality

$$\min\left(\frac{a}{a+b}, \frac{c}{c+d}\right) \geq \frac{\min(a, c)}{\min(a, c) + \max(b, d)}. \tag{3.1.9}$$

Let

$$X \equiv \min\left(\frac{a}{a+b}, \frac{c}{c+d}\right) - \frac{\min(a, c)}{\min(a, c) + \max(b, d)}.$$

If $a \geq c$, then, we obtain:

$$X = \min\left(\frac{a}{a+b}, \frac{c}{c+d}\right) - \frac{c}{c + \max(b,d)}.$$

If $\frac{a}{a+b} \geq \frac{c}{c+d}$, then,

$$X = \frac{c}{c+d} - \frac{c}{c + \max(b,d)} \geq 0.$$

If $\frac{a}{a+b} \leq \frac{c}{c+d}$, then, from (3.1.6),

$$X = \frac{a}{a+b} - \frac{c}{c + \max(b,d)} \geq \frac{c}{b+c} - \frac{c}{c + \max(b,d)} \geq 0.$$

Let $a < c$. Then,

$$X = \min\left(\frac{a}{a+b}, \frac{c}{c+d}\right) - \frac{a}{a + \max(b,d)}.$$

If $\frac{a}{a+b} \geq \frac{c}{c+d}$, then, from (1), we obtain

$$X = \frac{c}{c+d} - \frac{a}{a + \max(b,d)} \geq \frac{c}{c+d} - \frac{a}{a+d} \geq 0.$$

If $\frac{a}{a+b} \leq \frac{c}{c+d}$, then,

$$X = \frac{a}{a+b} - \frac{a}{a + \max(b,d)}.$$

In the same way, we can prove that

$$\frac{\max(b,d)}{\min(a,c) + \max(b,d)} \geq \left(\frac{b}{a+b}, \frac{d}{c+d}\right),$$

i.e., the first inequality is checked.

The second inequality in the theorem, as well as the Theorem 3.1.16 are proved by analogy.   □

Theorem 3.1.16 *For every predicate P:*

$$V(\bigcirc \exists x\, P(x)) \geq V(\exists x\, \bigcirc P(x)),$$

$$V(\bigcirc \forall x\, P(x)) \leq V(\forall x\, \bigcirc P(x)).$$

3.2 Extensions of the Intuitionistic Fuzzy Modal Operators

In this section, we introduce the first group of extended intuitionistic fuzzy modal operators.

First, by analogy with the IFS-operators from [7, 9], in the period 1988–1993, we define eight new modal operators.

Let A be a fixed formula for which $V(A) = \langle a, b \rangle$ and $\alpha, \beta, \gamma, \delta, \varepsilon, \eta \in [0, 1]$. We define operators D_α, $F_{\alpha,\beta}$, $G_{\alpha,\beta}$, $H_{\alpha,\beta}$, $H^*_{\alpha,\beta}$, $J_{\alpha,\beta}$, $J^*_{\alpha,\beta}$ and $X_{\alpha,\beta,\gamma,\delta,\varepsilon,\eta}$ by:

$$V(D_\alpha(A)) = \langle a + \alpha.(1 - a - b), b + (1 - \alpha).(1 - a - b) \rangle,$$
$$V(F_{\alpha,\beta}(A)) = \langle a + \alpha.(1 - a - b), b + \beta.(1 - a - b) \rangle, \text{ for } \alpha + \beta \le 1,$$
$$V(G_{\alpha,\beta}(A)) = \langle \alpha.a, \beta.b \rangle,$$
$$V(H_{\alpha,\beta}(A)) = \langle \alpha.a, b + \beta.(1 - a - b) \rangle,$$
$$V(H^*_{\alpha,\beta}(A)) = \langle \alpha.a, b + \beta.(1 - \alpha.a - b) \rangle,$$
$$V(J_{\alpha,\beta}(A)) = \langle a + \alpha.(1 - a - b), \beta.b \rangle,$$
$$V(J^*_{\alpha,\beta}(A)) = \langle a + \alpha.(1 - a - \beta.b), \beta.b \rangle,$$
$$V(X_{\alpha,\beta,\gamma,\delta,\varepsilon,\eta}(A)) = \langle \alpha.a + \beta.(1 - a - \gamma.b), \delta.b + \varepsilon.(1 - \eta.a - b) \rangle,$$

for

$$\alpha + \varepsilon - \varepsilon.\eta \le 1 \tag{3.2.1}$$

$$\beta + \delta - \beta\gamma \le 1, \tag{3.2.2}$$

$$\beta + \varepsilon \le 1. \tag{3.2.3}$$

In [10], it was mentioned that the third condition (3.2.3) was omitted in [7, 9]. It was introduced to the definition in [10], because without it, e.g., for constant $U^* = \langle 0, 0 \rangle$ we obtain

$$X_{0,1,0,0,1,0}(U^*) = \langle 1, 1 \rangle,$$

which is impossible.

Obviously,

$$\Box A = D_0(A),$$

$$\Diamond A = D_1(A),$$

$$D_\alpha(A) = F_{\alpha,1-\alpha}(A),$$

$$\Box A = X_{1,0,r,1,1,1}(A),$$

$$\Diamond A = X_{1,1,1,1,0,r}(A),$$

$$D_\alpha(A) = X_{1,\alpha,1,1,1-\alpha,1}(A),$$

$$F_{\alpha,\beta}(A) = X_{1,\alpha,1,1,\beta,1}(A), \text{ for } \alpha + \beta \leq 1,$$

$$G_{\alpha,\beta}(A) = X_{\alpha,0,r,\beta,0,r}(A),$$

$$H_{\alpha,\beta}(A) = X_{\alpha,0,r,1,\beta,1}(A),$$

$$H^*_{\alpha,\beta}(A) = X_{\alpha,0,r,\beta,0,\alpha}(A),$$

$$J_{\alpha,\beta}(A) = X_{1,\alpha,1,\beta,0,r}(A),$$

$$J^*_{\alpha,\beta}(A) = X_{1,\alpha,\beta,\beta,0,r}(A),$$

where r is an arbitrary real number in $[0, 1]$.
Let us define for every formula A:

$$V(D_\alpha(A)) = D_\alpha(V(A)),$$

$$V(F_{\alpha,\beta}(A)) = F_{\alpha,\beta}(V(A)),$$

$$V(G_{\alpha,\beta}(A)) = G_{\alpha,\beta}(V(A)),$$

$$V(H_{\alpha,\beta}(A)) = H_{\alpha,\beta}(V(A)),$$

$$V(H^*_{\alpha,\beta}(A)) = H^*_{\alpha,\beta}(V(A)),$$

$$V(J_{\alpha,\beta}(A)) = J_{\alpha,\beta}(V(A)),$$

$$V(J^*_{\alpha,\beta}(A)) = J^*_{\alpha,\beta}(V(A)),$$

$$V(X_{\alpha,\beta,\gamma,\delta,\varepsilon,\eta}(A)) = X_{\alpha,\beta,\gamma,\delta,\varepsilon,\eta}(V(A)).$$

To every formula A, the evaluation function V assigns for $D_\alpha(A)$ a point from the segment between $V(\Box A)$ and $V(\Diamond A)$ depending on the value of the argument $\alpha \in [0, 1]$ (see Fig. 3.6). As in the case of some of the above operations, this construction needs auxiliary elements which are shown in Fig. 3.6.

To every formula A, the evaluation function V assigns for $F_{\alpha,\beta}(A)$ a point from the triangle with vertices $V(A)$, $V(\Box A)$ and $V(\Diamond A)$, depending on the value of the arguments $\alpha, \beta \in [0, 1]$ for which $\alpha + \beta \leq 1$ (see Fig. 3.7).

To every formula A, the evaluation function V assigns for $G_{\alpha,\beta}(A)$ a point in the rectangle whose vertices are the point $V(A)$ and points with coordinates, $\langle pr_1 V(A), 0 \rangle$, $\langle 0, pr_2 V(A) \rangle$ and $\langle 0, 0 \rangle$, where $pr_i p$ is the i-th projection ($i = 1, 2$) (see Fig. 3.8).

Fig. 3.6 Second geometrical interpretation of operator D_α

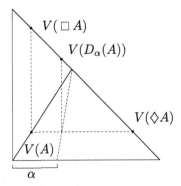

Fig. 3.7 Second geometrical interpretation of operator $F_{\alpha,\beta}$

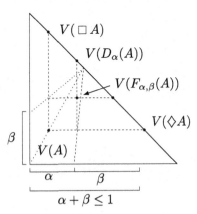

To every formula A, the evaluation function V assigns for $K_{\alpha,\beta}(A)$ a point $V(H_{\alpha,\beta}(A))$ from the rectangle whose vertices are the points with coordinates $\langle 0, pr_2V(A) \rangle$ and $\langle 0, pr_2V(\Box A) \rangle$ and vertices $V(\Box A)$ and $V(A)$, depending on the value of the parameters $\alpha, \beta \in [0, 1]$ (see Fig. 3.9).

To every formula A, the evaluation function V assigns for $J_{\alpha,\beta}(A)$ a point $V(J_{\alpha,\beta}(A))$ from the rectangle whose vertices are the points with coordinates $\langle pr_1V(\Diamond A), 0 \rangle$, $\langle pr_1V(A), 0 \rangle$ and vertices $V(A)$ and $V(\Diamond A)$, depending on the value of the parameters $\alpha, \beta \in [0, 1]$ (see Fig. 3.10).

To every formula A, the evaluation function V assigns for $H_{\alpha,\beta}^*(A)$ a point $V(H_{\alpha,\beta}^*(A))$ from the trapezoid with vertices with coordinates $\langle 0, pr_2V(A) \rangle$ and $\langle 0, 1 \rangle$ and vertices $V(\Box A)$ and $V(A)$, depending on the value of the parameters $\alpha, \beta \in [0, 1]$ (see Fig. 3.11).

To every formula A, the evaluation function V assigns for $J_{\alpha,\beta}^*(A)$ a point $V(J_{\alpha,\beta}^*(A))$ from the trapezoid with vertices with coordinates $\langle 1, 0 \rangle$ and $\langle pr_1V(A), 0 \rangle$ and vertices $V(A)$ and $V(\Diamond A)$, depending on the value of the parameters $\alpha, \beta \in [0, 1]$ (see Fig. 3.12).

Fig. 3.8 Second geometrical
interpretation of operator $G_{\alpha,\beta}$

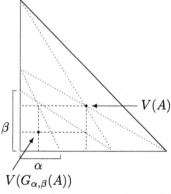

$V(G_{\alpha,\beta}(A))$

Fig. 3.9 Second geometrical
interpretation of operator $H_{\alpha,\beta}$

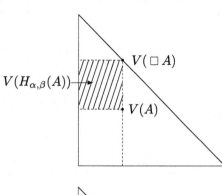

Fig. 3.10 Second geometrical
interpretation of operator $J_{\alpha,\beta}$

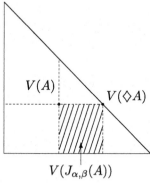

$V(J_{\alpha,\beta}(A))$

Below, we formulate and prove assertions, that by the moment have been checked
only for some intuitionistic fuzzy implications and negation, which gives rise to the
following important open problem for solving in future.

Open Problem 14 Check these assertions for all the remaining implications and
negations.

Fig. 3.11 Second geometrical interpretation of operator $H^*_{\alpha,\beta}$

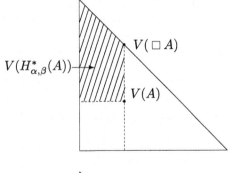

Fig. 3.12 Second geometrical interpretation of operator $J^*_{\alpha,\beta}$

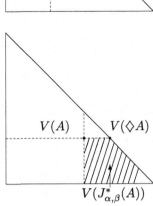

Theorem 3.2.1 *If formula A is a tautology, then:*

(a) *for every two* $\alpha, \beta \in [0, 1]$: $D_\alpha(A)$, $F_{\alpha,\beta}(A)$, *for* $\alpha + \beta \leq 1$, $J_{\alpha,\beta}(A)$ *and* $J^*_{\alpha,\beta}(A)$ *are tautologies,*
(b) $G_{\alpha,\beta}(A)$, $H_{\alpha,\beta}(A)$, $H^*_{\alpha,\beta}(A)$, *are IFTs,*
(c) *for every* $\alpha, \varepsilon, \eta$ *so that* $\alpha \geq \varepsilon(1 - \eta)$, $X_{\alpha,\beta,\gamma,\delta,\varepsilon,\eta}(A)$ *is an IFT.*

Proof (b) From $V(A) = \langle 1, 0 \rangle$ it follows that

$$V(G_{\alpha,\beta}(A)) = \langle \alpha, 0 \rangle,$$

$$V(H_{\alpha,\beta}(A)) = V(H^*_{\alpha,\beta}(A)) = \langle \alpha, \beta(1 - 1 - 0) \rangle = \langle \alpha, 0 \rangle,$$

i.e., $G_{\alpha,\beta}(A)$, $H_{\alpha,\beta}(A)$ and $H^*_{\alpha,\beta}(A)$ are IFTs. The other cases are proved by analogy. □

Theorem 3.2.2 *If formula A is an IFT, then:*

(a) *for* $\alpha \geq 0.5$ $D_\alpha(A)$ *is an IFT,*
(b) *for* $\alpha \geq \beta$ *and* $\alpha + \beta \leq 1$ $F_{\alpha,\beta}(A)$ *is an IFT,*
(c) *for* $\alpha \geq \beta$ $G_{\alpha,\beta}(A)$ *is an IFT,*
(d) $J_{\alpha,\beta}(A)$, $J^*_{\alpha,\beta}(A)$ *are IFTs,*
(e) *for* $\alpha \geq \delta$, $\beta \geq \varepsilon$, $\eta \geq \gamma$ $X_{\alpha,\beta,\gamma,\delta,\varepsilon,\eta}(A)$ *is an IFT.*

It can be easily seen that for every $a, b, \alpha, \beta, \gamma, \delta, \varepsilon, \eta \in [0, 1]$ and $a + b \leq 1$, if

$$O \in \{D_\alpha, F_{\alpha,\beta}, G_{\alpha,\beta}, H_{\alpha,\beta}, H^*_{\alpha,\beta}, J_{\alpha,\beta}, J^*_{\alpha,\beta}, X_{\alpha,\beta,\gamma,\delta,\varepsilon,\eta}\},$$

then

$$O(\langle a, b \rangle) \to_4 \langle a, b \rangle \quad \text{and} \quad O(\langle a, b \rangle) \to_{11} \langle a, b \rangle$$

are IFTs. For example,

$$
\begin{aligned}
&V(F_{\alpha,\beta}(\langle a, b \rangle) \to_{11} \langle a, b \rangle) \\
&= \langle a + \alpha.(1 - a - b), b + \beta(1 - a - b) \rangle \to_{11} \langle a, b \rangle \\
&= \langle 1 - (1 - a).\mathrm{sg}(\alpha(1 - a - b)), b.\mathrm{sg}(\alpha(1 - a - b)).\mathrm{sg}(b + \beta.(1 - a - b) - b) \rangle, \\
&= \langle 1 - (1 - a).\mathrm{sg}(\alpha(1 - a - b)), b.\mathrm{sg}(\alpha(1 - a - b)).\mathrm{sg}(\beta.(1 - a - b)) \rangle
\end{aligned}
$$

$$
= \begin{cases}
\langle 1, 0 \rangle, & \text{if } \alpha = 0 \text{ or } a + b = 1 \\
\langle a, b \rangle, & \text{if } \alpha > 0 \text{ and } a + b < 1
\end{cases},
$$

i.e., $F_{\alpha,\beta}(\langle a, b \rangle) \to_{11} \langle a, b \rangle$ is an IFT.

On the other hand, we can see that there are cases in which for the different implications, different conditions must hold for the validity of some expression. The following two assertions serve as examples.

Theorem 3.2.3 *For every* $a, b, \alpha, \beta, \gamma, \delta, \varepsilon, \eta, \alpha', \beta', \gamma', \delta', \varepsilon', \eta' \in [0, 1]$ *and* $a + b \leq 1$, *if* $O \in \{D, F, G, H, H^*, J, J^*\}$, *then:*

(a) $O_{\alpha,\beta}(\langle a, b \rangle) \to_{11} O_{\alpha',\beta'}(\langle a, b \rangle)$ *is a tautology for* $\alpha \leq \alpha'$,

(b) $X_{\alpha,\beta,\gamma,\delta,\varepsilon,\eta}(\langle a, b \rangle) \to_{11} X_{\alpha',\beta',\gamma',\delta',\varepsilon',\eta'}(\langle a, b \rangle)$ *is a tautology for* $\alpha \leq \alpha', \beta \leq \beta'$ *and* $\beta\gamma \geq \beta'\gamma'$.

Proof (b) Let $\alpha \leq \alpha', \beta \leq \beta'$ and $\beta.\gamma \geq \beta'.\gamma'$. Then,

$$
\begin{aligned}
&V(X_{\alpha,\beta,\gamma,\delta,\varepsilon,\eta}(\langle a, b \rangle) \to_{11} X_{\alpha',\beta',\gamma',\delta',\varepsilon',\eta'}(\langle a, b \rangle)) \\
&= \langle \alpha a + \beta(1 - a - \gamma b), \delta b + \varepsilon(1 - \eta a - b) \rangle \\
&\quad \to_{11} \langle \alpha' a + \beta'(1 - a - \gamma' b), \delta' b + \varepsilon'(1 - \eta' a - b) \rangle \\
&= \langle 1 - (1 - (\alpha' a + \beta'(1 - a - \gamma' b)))\mathrm{sg}(\alpha a + \beta(1 - a - \gamma b) \\
&\quad - (\alpha' a + \beta'(1 - a - \gamma' b))), (\delta' b + \varepsilon'(1 - \eta' a - b)) \\
&\quad .\mathrm{sg}(\alpha a + \beta(1 - a - \gamma b) - (\alpha' a + \beta'(1 - a - \gamma' b))) \\
&\quad .\mathrm{sg}(\delta' b + \varepsilon'(1 - \eta' a - b) - (\delta b + \varepsilon(1 - \eta a - b))) \rangle \\
&= \langle 1, 0 \rangle,
\end{aligned}
$$

because

$$
\begin{aligned}
&\alpha a + \beta(1 - a - \gamma b) - (\alpha' a + \beta'(1 - a - \gamma' b)) \\
&= a(\alpha - \alpha') + (1 - a)(\beta - \beta') - b(\beta\gamma - \beta'\gamma') \leq 0
\end{aligned}
$$

and therefore, $sg(\alpha a + \beta(1 - a - \gamma b) - (\alpha' a + \beta'(1 - a - \gamma' b)) = 0$. The other cases are proved similarly.   □

Theorem 3.2.4 *For every* $a, b, \alpha, \beta, \gamma, \delta, \varepsilon, \eta, \alpha', \beta', \gamma', \delta', \varepsilon', \eta' \in [0, 1]$ *and* $a + b \leq 1$, *if* $O \in \{D, F, G, H, H^*, J, J^*\}$, *then*

(a) $O_{\alpha,\beta}(\langle a, b \rangle) \rightarrow_4 O_{\alpha',\beta'}(\langle a, b \rangle)$ *is an IFT for* $\alpha \leq \alpha'$ *or* $\beta \geq \beta'$,

(b) $X_{\alpha,\beta,\gamma,\delta,\varepsilon,\eta}(\langle a, b \rangle) \rightarrow_4 X_{\alpha',\beta',\gamma',\delta',\varepsilon',\eta'}(\langle a, b \rangle)$ *is an IFT for* $\alpha \leq \alpha'$, $\beta \leq \beta'$ *and* $\gamma \geq \gamma'$ *or for* $\delta \geq \delta'$, $\varepsilon \geq \varepsilon'$ *and* $\eta \leq \eta'$.

Open Problem 15 Which other implications satisfy Theorems 3.2.3 and 3.2.4?

There are properties that, probably, are specific for $\neg_1$. For example, the following assertion is valid for $\neg_1$, but for $0 < \varepsilon \leq \eta \leq 1$ it is not valid for $\neg_{45,\varepsilon,\eta}$ (by definition, $\neg_1$ coincides with $\neg_{45,0,0}$).

Theorem 3.2.5 *For every* $a, b, \alpha, \beta \in [0, 1]$ *and* $a + b \leq 1$:

(a) $V(F_{\alpha,\beta}(\langle a, b \rangle)) = V(\neg_1 F_{\beta,\alpha}(\neg_1 \langle a, b \rangle))$, *for* $\alpha + \beta \leq 1$,
(b) $V(G_{\alpha,\beta}(\langle a, b \rangle)) = V(\neg_1 G_{\beta,\alpha}(\neg_1 \langle a, b \rangle))$,
(c) $V(H_{\alpha,\beta}(\langle a, b \rangle)) = V(\neg_1 J_{\beta,\alpha}(\neg_1 \langle a, b \rangle))$,
(d) $V(J_{\alpha,\beta}(\langle a, b \rangle)) = V(\neg_1 H_{\beta,\alpha}(\neg_1 \langle a, b \rangle))$,
(e) $V(H^*_{\alpha,\beta}(\langle a, b \rangle)) = V(\neg_1 J^*_{\beta,\alpha}(\neg_1 \langle a, b \rangle))$,
(f) $V(J^*_{\alpha,\beta}(\langle a, b \rangle)) = V(\neg_1 H^*_{\beta,\alpha}(\neg_1 \langle a, b \rangle))$.

Open Problem 16 Which other negations satisfy Theorems 3.2.5?

Let for every two formulas A and B: $A \leftarrow B$ iff $B \rightarrow A$.

Theorem 3.2.6 *For every two formulas* A *and* B, *for every two real numbers* $\alpha, \beta \in [0, 1]$ *and for implication* $\rightarrow_4$:

(a) $F_{\alpha,\beta}(A \wedge B) \rightarrow_4 F_{\alpha,\beta}(A) \wedge F_{\alpha,\beta}(B)$, *for* $\alpha + \beta \leq 1$,
(b) $F_{\alpha,\beta}(A \vee B) \leftarrow_4 F_{\alpha,\beta}(A) \vee F_{\alpha,\beta}(B)$, *for* $\alpha + \beta \leq 1$,
(c) $G_{\alpha,\beta}(A \wedge B) = G_{\alpha,\beta}(A) \wedge G_{\alpha,\beta}(B)$,
(d) $G_{\alpha,\beta}(A \vee B) = G_{\alpha,\beta}(A) \vee G_{\alpha,\beta}(B)$,
(e) $H_{\alpha,\beta}(A \wedge B) \rightarrow_4 H_{\alpha,\beta}(A) \wedge H_{\alpha,\beta}(B)$,
(f) $H_{\alpha,\beta}(A \vee B) \leftarrow_4 H_{\alpha,\beta}(A) \vee H_{\alpha,\beta}(B)$,
(g) $J_{\alpha,\beta}(A \wedge B) \leftarrow_4 H_{\alpha,\beta}(A) \wedge H_{\alpha,\beta}(B)$,
(h) $H_{\alpha,\beta}(A \vee B) \rightarrow_4 H_{\alpha,\beta}(A) \vee H_{\alpha,\beta}(B)$,
(i) $H^*_{\alpha,\beta}(A \wedge B) \rightarrow_4 H^*_{\alpha,\beta}(A) \wedge H^*_{\alpha,\beta}(B)$,
(j) $H^*_{\alpha,\beta}(A \vee B) \leftarrow_4 H^*_{\alpha,\beta}(A) \vee H^*_{\alpha,\beta}(B)$,
(k) $J^*_{\alpha,\beta}(A \wedge B) \leftarrow_4 H^*_{\alpha,\beta}(A) \wedge H^*_{\alpha,\beta}(B)$,
(l) $H^*_{\alpha,\beta}(A \vee B) \rightarrow_4 H^*_{\alpha,\beta}(A) \vee H^*_{\alpha,\beta}(B)$

are IFTs.

Proof (a) For the formulas A and B:

$$V(F_{\alpha,\beta}(A \wedge B) \rightarrow_4 F_{\alpha,\beta}(A) \wedge F_{\alpha,\beta}(B))$$
$$= \langle \min(a,c) + \alpha(1 - \min(a,c) - \max(b,d)), \max(b,d) +$$
$$\beta(1 - \min(a,c) - \max(b,d)) \rangle \rightarrow \langle \min(a + \alpha(1 - a - b), c$$
$$+ \alpha(1 - c - d)), \max(b + \beta(1 - a - b), d + \beta(1 - c - d)) \rangle$$
$$= \langle \max(\max(b,d) + \beta(1 - \min(a,c) - \max(b,d)), \min(a +$$
$$\alpha(1 - a - b), c + \alpha(1 - c - d))), \min(\min(a,c) + \alpha(1 - \min(a,c)$$
$$- \max(b,d)), \max(b + \beta(1 - a - b), d + \beta(1 - c - d)))) \rangle$$

and

$$\max(\max(b,d) + \beta(1 - \min(a,c) - \max(b,d)), \min(a +$$
$$\alpha(1 - a - b), c + \alpha(1 - c - d))) - \min(\min(a,c) + \alpha(1 -$$
$$\min(a,c) - \max(b,d)), \max(b + \beta(1 - a - b), d + \beta(1 - c - d)))$$
$$\geq \max(b,d) + \beta(1 - \min(a,c) - \max(b,d))$$
$$- \max(b + \beta(1 - a - b), d + \beta(1 - c - d)) \geq 0,$$

i.e.,

$$F_{\alpha,\beta}(A \wedge B) \rightarrow_4 F_{\alpha,\beta}(A) \wedge F_{\alpha,\beta}(B)$$

is an IFT.

Formulas (b)–(l) are checked by analogy. □

Open Problem 17 Which other conjunctions and disjunctions (whenever be defined, following the ideas from Sect. 1.7) have similar properties?

Theorem 3.2.7 *For every predicate A, and for every $\alpha, \beta \in [0,1]$, such that $\alpha + \beta \leq 1$:*

(a) $V(\exists x F_{\alpha,\beta}(P(x))) \leq V(F_{\alpha,\beta} \exists x P(x))$,
(b) $V(\forall x F_{\alpha,\beta}(P(x))) \geq V(F_{\alpha,\beta} \forall x P(x))$.

Corollary 3.2.1 *For every predicate A, and for every $\alpha, \beta \in [0,1]$, such that $\alpha + \beta \leq 1$:*

(a) $V(\exists x D_\alpha(P(x))) \leq V(D_\alpha \exists x P(x))$,
(b) $V(\forall x D_\alpha(P(x))) \geq V(D_\alpha \forall x P(x))$.

Theorem 3.2.8 *For every formula A and for every $\alpha, \beta, \gamma, \delta \in [0,1]$ such that $\alpha + \beta \leq 1$ and $\gamma + \delta \leq 1$:*

$$V(F_{\alpha,\beta}(F_{\gamma,\delta}(A))) = V(F_{\alpha+\gamma-\alpha\gamma-\alpha\delta,\beta+\delta-\beta\gamma-\beta\delta}(A)).$$

Proof Let for $\alpha, \beta, \gamma, \delta \in [0,1]$: $\alpha + \beta \leq 1$ and $\gamma + \delta \leq 1$. Let $V(A) = \langle a, b \rangle$. Then,

$$V(F_{\alpha,\beta}(F_{\gamma,\delta}(A)))$$
$$= F_{\alpha,\beta}(\langle a + \gamma(1 - a - b), b + \delta(1 - a - b)\rangle)$$
$$= \langle a + \gamma(1 - a - b) + \alpha(1 - a - \gamma(1 - a - b) - b - \delta(1 - a - b)),$$
$$\quad b + \delta(1 - a - b) + \beta(1 - a - \gamma(1 - a - b) - b - \delta(1 - a - b))\rangle$$
$$= \langle a + (\alpha + \gamma - \alpha\gamma - \alpha\delta)(1 - a - b),$$
$$\quad b + (\beta + \delta - \beta\gamma - \beta\delta)(1 - a - b)\rangle$$
$$= V(F_{\alpha+\gamma-\alpha\gamma-\alpha\delta,\beta+\delta-\beta\gamma-\beta\delta}(A)).$$

This completes the proof. □

Corollary 3.2.2 *For every predicate A, and for every $\alpha \in [0, 1]$:*

(a) $V(D_\alpha(D_\beta(A))) = V(D_\beta(A))$,
(b) $V(D_\alpha(F_{\beta,\gamma}(A))) = V(D_{\alpha+\beta-\alpha\beta-\alpha\gamma}(A))$, *for $\beta + \gamma \leq 1$,*
(c) $V(F_{\alpha,\beta}(D_\gamma(A))) = V(D_\gamma(A))$.

Theorem 3.2.9 *For every formula A and for every $\alpha, \beta, \gamma, \delta \in [0, 1]$:*

$$V(G_{\alpha,\beta}(G_{\gamma,\delta}(A))) = V(G_{\alpha\gamma,\beta\delta}(A)).$$

Theorem 3.2.10 *For every formula A and for every $\alpha, \beta \in [0, 1]$:*

(a) $V(\square\, F_{\alpha,\beta}(A)) \geq V(F_{\alpha,\beta}\,\square\, A)$ *for $\alpha + \beta \leq 1$,*
(b) $V(\Diamond\, F_{\alpha,\beta}(A)) \leq V(F_{\alpha,\beta}\Diamond\, A)$ *for $\alpha + \beta \leq 1$,*

(c) $V(\square\, G_{\alpha,\beta}(A)) \leq V(G_{\alpha,\beta}\,\square\, A)$,
(d) $V(\Diamond\, G_{\alpha,\beta}(A)) \geq V(G_{\alpha,\beta}\Diamond\, A)$,

(e) $V(\square\, H_{\alpha,\beta}(A)) \leq V(H_{\alpha,\beta}\,\square\, A)$,
(f) $V(\Diamond\, H_{\alpha,\beta}(A)) \geq V(H_{\alpha,\beta}\Diamond\, A)$,

(g) $V(\square\, J_{\alpha,\beta}(A)) \geq V(J_{\alpha,\beta}\,\square\, A)$,
(h) $V(\Diamond\, J_{\alpha,\beta}(A)) \leq V(J_{\alpha,\beta}\Diamond\, A)$,

(i) $V(\square\, H^*_{\alpha,\beta}(A)) \leq V(H^*_{\alpha,\beta}\,\square\, A)$,
(j) $V(\Diamond\, J^*_{\alpha,\beta}(A)) \geq V(J^*_{\alpha,\beta}\Diamond\, A)$.

Theorem 3.2.11 *For every A and for every $\alpha, \beta, \gamma, \delta \in [0, 1]$:*

$$G_{\alpha,\beta}(G_{\gamma,\delta}(A)) = G_{\alpha\gamma,\beta\delta}(A).$$

Theorem 3.2.12 *For every formula A and for every $\alpha, \beta \in [0, 1]$:*

(a) $V(H_{\alpha,\beta}(G_{\gamma,\delta}(A))) \leq V(G_{\gamma,\delta}(H_{\alpha,\beta}(A)))$,
(b) $V(J_{\alpha,\beta}(G_{\gamma,\delta}(A))) \leq V(G_{\gamma,\delta}(J_{\alpha,\beta}(A)))$,
(c) $V(H^*_{\alpha,\beta}(G_{\gamma,\delta}(A))) \leq V(G_{\gamma,\delta}(H^*_{\alpha,\beta}(A)))$,
(d) $V(J^*_{\alpha,\beta}(G_{\gamma,\delta}(A))) \leq V(G_{\gamma,\delta}(J^*_{\alpha,\beta}(A)))$.

Theorem 3.2.13 *Let A be a formula and x be a variable. Then, for every $\alpha, \beta \in$ [0, 1]:*

$$V(\forall x G_{\alpha,\beta}(A)) = V(G_{\alpha,\beta}(\forall x A)),$$

$$V(\exists x G_{\alpha,\beta}(A)) = V(G_{\alpha,\beta}(\exists x A)).$$

We finish with an assertion for the operator X, following [11].

Theorem 3.2.14 *For every two IFPs $\langle \mu, \nu \rangle$ and $\langle \rho, \sigma \rangle$, there are real numbers $a, b, c, d, e, f \in [0, 1]$ satisfying (3.2.1)–(3.2.3), such that*

$$V(X_{a,b,c,d,e,f}(\langle \mu, \nu \rangle)) = \langle \rho, \sigma \rangle.$$

Proof Let $\mu, \nu, \rho, \sigma \in [0, 1]$, such that $\mu + \nu \leq 1, \rho + \sigma \leq 1$. We search for $a, b, c, d, e, f \in [0, 1]$ that satisfy (3.2.1)–(3.2.3) and for which

$$\langle \rho, \sigma \rangle = V(X_{a,b,c,d,e,f}(\langle \mu, \nu, \rangle)) = \langle a\mu + b(1 - \mu - c\nu), d\nu + e(1 - f\mu - \nu) \rangle,$$

i.e.,

$$\rho = a\mu + b(1 - \mu - c\nu),$$

$$\sigma = d\nu + e(1 - f\mu - \nu).$$

We discuss nine cases.

Case 1. $\pi = \mu = 0$. Then, $\nu = 1$. We put

$$a = c = e = f = 0, \ b = \rho, \ d = \sigma.$$

Then, conditions (3.2.1)–(3.2.3) are valid and

$$X_{0,\rho,0,\sigma,0,0}(\langle \mu, \nu, \rangle) = \langle 0 + \rho(1 - 0 - 0 \times 1), \sigma \times 1 + 0(1 - 0 \times \mu - 1) \rangle = \langle \rho, \sigma \rangle.$$

Case 2. $\pi = \nu = 0$. Then, $\mu = 1$. We put

$$a = \rho, \ b = c = d = f = 0, \ e = \sigma.$$

Then, conditions (3.2.1)–(3.2.3) are valid and

$$X_{\rho,0,0,0,\sigma,0}(\langle \mu, \nu, \rangle) = \langle \rho + 0 \times (1 - 0 \times 1 - 0), 0 + \sigma(1 - 0 \times 1 - 1) \rangle = \langle \rho, \sigma \rangle.$$

When $\pi = 0$ and $\mu, \nu > 0$, there are three (sub)cases. It is important to mention that now $\mu, \nu < 1$.

Case 3. $\rho > \mu$. Then, from $\pi = 0$ it follows that $\mu = 1 - \nu$, and hence $\sigma \leq 1 - \rho < 1 - \mu = \nu$. So, we put

$$a = 1, \ b = \frac{\rho - \mu}{1 - \mu}, c = e = f = 0, \ d = \frac{\sigma}{\nu}.$$

Then, conditions (3.2.1)–(3.2.3) are valid and

$$X_{1,\frac{\rho-\mu}{1-\mu},0,\frac{\sigma}{\nu},0,0}(\langle\mu,\nu,\rangle) = \left\langle \mu + \frac{\rho-\mu}{1-\mu}(1-\mu), \frac{\sigma}{\nu}\nu \right\rangle = \langle\rho,\sigma\rangle.$$

Case 4. $\sigma > \nu$. Then, from $\pi = 0$ it follows again that $\mu = 1 - \nu$, and hence $\rho \le 1 - \sigma < 1 - \nu = \mu$. So, we put

$$a = \frac{\rho}{\mu},\ b = c = f = 0,\ d = 1, e = \frac{\sigma-\nu}{1-\nu}.$$

Then, conditions (3.2.1)–(3.2.3) are valid and

$$X_{\frac{\rho}{\mu},0,0,1,\frac{\sigma-\nu}{1-\nu},0,0}(\langle\mu,\nu,\rangle) = \left\langle \frac{\rho}{\mu}\mu + 0, \nu + \frac{\sigma-\nu}{1-\nu}(1-\nu) \right\rangle = \langle\rho,\sigma\rangle.$$

Case 5. $\rho \le \mu$ and $\sigma \le \nu$. Then, we put

$$a = \frac{\rho}{\mu},\ b = c = e = f = 0,\ d = \frac{\sigma}{\nu}.$$

Then, conditions (3.2.1)–(3.2.3) are valid and

$$X_{\frac{\rho}{\mu},0,0,\frac{\sigma}{\nu},0,0}(\langle\mu,\nu,\rangle) = \left\langle \frac{\rho}{\mu}\mu + 0, \frac{\sigma}{\nu}\nu + 0 \right\rangle = \langle\rho,\sigma\rangle.$$

When $\pi > 0$, then, $\mu, \nu < 1$.

Case 6. $\rho > \mu$ and $\sigma > \nu$. Then, we put

$$a = c = d = f = 1,\ b = \frac{\rho-\mu}{\pi},\ e = \frac{\sigma-\nu}{\pi}.$$

Then, conditions (3.2.1)–(3.2.3) are valid, because:

$$a + e - e.f = 1 + e - e = 1 \le 1,$$

$$b + d - b.c = d = \frac{\sigma-\nu}{\pi} \le 1,$$

$$b + e = \frac{\rho-\mu}{\pi} + \frac{\sigma-\nu}{\pi} = \frac{\rho+\sigma-\mu-\nu}{\pi} \le 1.$$

All other checks are done in a similar way. Now,

$$X_{1,\frac{\rho-\mu}{\pi},1,1,\frac{\sigma-\nu}{\pi},1}(\langle\mu,\nu,\rangle) = \left\langle \mu + \frac{\rho-\mu}{\pi}\pi, \nu + \frac{\sigma-\nu}{\pi}\pi \right\rangle = \langle\rho,\sigma\rangle.$$

Case 7. $\rho > \mu$ and $\sigma \leq \nu$. Then, as in Case 3, we put

$$a = 1, \ b = \frac{\rho - \mu}{1 - \mu}, c = e = f = 0, \ d = \frac{\sigma}{\nu}.$$

Then, conditions (3.2.1)–(3.2.3) are valid and

$$X_{1,\frac{\rho-\mu}{1-\mu},0,\frac{\sigma}{\nu},0,0}(\langle \mu, \nu, \rangle) = \left\langle \mu + \frac{\rho - \mu}{1 - \mu}(1 - \mu), \frac{\sigma}{\nu}\nu \right\rangle = \langle \rho, \sigma \rangle.$$

Case 8. $\rho \leq \mu$ and $\sigma > \nu$. Then, as in point 4, we put

$$a = \frac{\rho}{\mu}, \ b = c = f = 0, \ d = 1, e = \frac{\sigma - \nu}{1 - \nu}.$$

Then, conditions (3.2.1)–(3.2.3) are valid and

$$X_{\frac{\rho}{\mu},0,0,1,\frac{\sigma-\nu}{1-\nu},0,0}(\langle \mu, \nu, \rangle) = \left\langle \frac{\rho}{\mu}\mu + 0, \nu + \frac{\sigma - \nu}{1 - \nu}(1 - \nu) \right\rangle = \langle \rho, \sigma \rangle.$$

Case 9. $\rho \leq \mu$ and $\sigma \leq \nu$. Then, as in Case 5, we put

$$a = \frac{\rho}{\mu}, \ b = c = e = f = 0, \ d = \frac{\sigma}{\nu}.$$

Then, conditions (3.2.1)–(3.2.3) are valid and

$$X_{\frac{\rho}{\mu},0,0,\frac{\sigma}{\nu},0,0}(\langle \mu, \nu, \rangle) = \left\langle \frac{\rho}{\mu}\mu + 0, \frac{\sigma}{\nu}\nu + 0 \right\rangle = \langle \rho, \sigma \rangle.$$

Therefore, the theorem is proved.   □

A modification of the above theorem is the following theorem.

Theorem 3.2.15 *For every two formulas A, B there exists an operator* $Y \in \{F_{\alpha,\beta},$ $G_{\alpha,\beta}, H_{\alpha,\beta}, H^*_{\alpha,\beta}, J_{\alpha,\beta}, J^*_{\alpha,\beta}\}$ *and there exist real numbers* $\alpha, \beta \in [0, 1]$ *such that*

$$V(A) = V(Y_{\alpha,\beta}(B)).$$

Proof Let everywhere $V(A) = \langle a, b \rangle$ and $V(B) = \langle c, d \rangle$, where $a, b, c, d \in [0, 1]$ and $a + b \leq 1$ and $c + d \leq 1$.
   The following 9 cases are possible for a, b, c and d.

Case 1. $a = c$ Then, for Y :
             and   if $Y = F$, then, $\alpha = \beta = 0$;
             $b = d$ if $Y = G$, then, $\alpha = \beta = 1$;
                       if $Y = H$ or $Y = H^*$, then, $\alpha = 1$ and $\beta = 0$;
                       if $Y = J$ or $Y = J^*$ then $\alpha = 0$ and $\beta = 1$.

Case 2. $a > c$ Then, $Y = F$ and $\alpha = \dfrac{a - c}{1 - c - d}$ and $\beta = 0$

and   (we shall note that $1 - c - d > 1 - a - b \geq 0$)
$b = d$ or $Y = J$ or $Y = J^*$, and α has the above
form and $\beta = 1$.

Case 3. $a < c$ Then, $Y = G$ and $\alpha = \dfrac{a}{c}$ and $\beta = 1$ (we note

and $that\, c > a \geq 0$)
$b = d$ or $Y = J$ or $Y = J^*$, and α has the above
form and $\beta = 0$.

Case 4. $a = c$ $Then,\, Y = F$ and $\alpha = 0$ and $\beta = \dfrac{b - d}{1 - c - d}$

and   (we note that $1 - c - d > 1 - a - b \geq 0$)
$b > d$ or $Y = H$ or $Y = H^*$, and $\alpha = 1$ and β has
the above form.

Case 5. $a > c$ Then, $Y = F$ and $\alpha = \dfrac{a - c}{1 - c - d}$ and

and $\beta = \dfrac{b - d}{1 - c - d}$ (we note that
$b > d$ $1 - c - d > 1 - a - b \geq 0$)

Case 6. $a < c$ Then, there are two subcases:
and
$b > d$.

6.1. $if\, b \leq 1 - c$, then, $Y = H$ and $\alpha = \dfrac{a}{c}$ and

$\beta = \dfrac{b - d}{1 - c - d}$

or $Y = H^*$ and $\alpha = \dfrac{a}{c}$ and $\beta = \dfrac{b - d}{1 - a - d}$

(we note that $1 - a - d > 1 - c - d \geq b - d > 0$
and $c > a \geq 0$)

6.2. $if\, b > 1 - c$, then, $Y = H^*$ and $\alpha = \dfrac{a}{c}$ and

$\beta = \dfrac{b - d}{1 - a - d}$ (we note that $1 - a - d \geq b - d > 0$

and $c > a \geq 0$)

Case 7. $a = c$ Then, $Y = G$ and $\alpha = 1$ and $\beta = \dfrac{b}{d}$ (we note

and   that $d > b \geq 0$)
$b < d$.
or $Y = J$ or $Y = H^*$, and $\alpha = 0$ and β has the above form.

Case 8. $a > c$ Then, there are two subcases:
     and
     $b < d$.

    8.1. $if a \leq 1 - d$, then, $Y = J$ and $\alpha = \dfrac{a - c}{1 - c - d}$ and

$$\beta = \frac{b}{d}$$

or $Y = J^*$ and $\alpha = \dfrac{a - c}{1 - b - c}$ and $\beta = \dfrac{b}{d}$ (we

note that $1 - c - b > 1 - c - d \geq a - c > 0$
and $d > b \geq 0$)

    8.2. $if a > 1 - d$, then, $Y = J^*$ and $\alpha = \dfrac{a - c}{1 - b - c}$

and $\beta = \dfrac{b}{d}$ (we note that $1 - c - b \geq a - c > 0$

and $d > b \geq 0$)

Case 9. $a < c$ Then, $Y = G$ and $\alpha = \dfrac{a}{c}$ and $\beta = \dfrac{b}{d}$ (we shall

    and   note that $c > a \geq 0$ and $d > b \geq 0$).
    $b < d$.

This completes the proof.   □

3.3 Second Type of Intuitionistic Fuzzy Modal Operators

In this section, following [7, 9, 12–17], several different modal-like operators of second type are defined and the consequences of their generalizations are discussed. We formulate the properties of these operators which hold for them but do not hold for their extensions.

The following two operators of modal type are similar to the operators in Sect. 3.1. Let for formula A: $V(A) = \langle \mu, \nu \rangle$. Then,

$$V(\boxplus A) = \left\langle \frac{\mu}{2}, \frac{\nu + 1}{2} \right\rangle, \tag{3.3.1}$$

$$V(\boxtimes A) = \left\langle \frac{\mu + 1}{2}, \frac{\nu}{2} \right\rangle. \tag{3.3.2}$$

All of their properties are valid for their first extensions. For a given real number $\alpha \in [0, 1]$ and formula A,

$$V(\boxplus_\alpha A) = \langle \alpha\mu, \alpha\nu + 1 - \alpha \rangle, \tag{3.3.3}$$

$$V(\boxtimes_\alpha A) = \langle \alpha\mu + 1 - \alpha, \alpha\nu \rangle. \tag{3.3.4}$$

Fig. 3.13 Second geometrical interpretation of operator $\boxplus_\alpha$

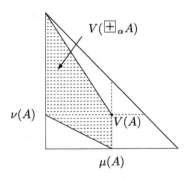

Fig. 3.14 Second geometrical interpretation of operator $\boxtimes_\alpha$

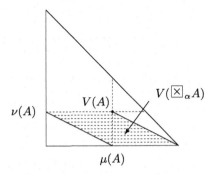

Obviously,

$$0 \le \alpha\mu + \alpha\nu + 1 - \alpha = 1 - \alpha(1 - \mu - \alpha\nu) \le 1.$$

For every formula A,

$$V(\boxplus_{0.5}A) = V(\boxplus A),$$

$$V(\boxtimes_{0.5}A) = V(\boxtimes A).$$

Therefore, the new operators "$\boxplus_\alpha$" and "$\boxtimes_\alpha$" are generalizations of $\boxplus$ and $\boxtimes$, respectively. Their geometrical interpretations are given in Figs. 3.13 and 3.14, respectively.

For every formula A, for every $\alpha \in [0, 1]$:

(a) $V(\boxplus_\alpha A) \le V(A) \le V(\boxtimes_\alpha A)$,

(b) $V(\neg_1\boxplus_\alpha\neg_1 A) = V(\boxtimes_\alpha A)$,

(c) $V(\boxplus_\alpha\boxplus_\alpha A) \le V(\boxplus_\alpha A)$,

(d) $V(\boxtimes_\alpha\boxtimes_\alpha A) \le V(\boxtimes_\alpha A)$.

For every formula A, and for every two real numbers $\alpha, \beta \in [0, 1]$,

(a) $V(\boxplus_\alpha \boxminus_\beta A) = V(\boxminus_\beta \boxplus_\alpha A)$,

(b) $V(\boxtimes_\alpha \boxtimes_\beta A) = V(\boxtimes_\beta \boxtimes_\alpha A)$,

(c) $V(\boxtimes_\alpha \boxminus_\beta A) \geq V(\boxplus_\beta \boxtimes_\alpha A)$.

For every formula A, and for every three real numbers $\alpha, \beta, \gamma \in [0, 1]$,

(a) $V(\boxplus_\alpha D_\beta(A)) = V(D_\beta(\boxplus_\alpha A))$,

(b) $V(\boxplus_\alpha F_{\beta,\gamma}(A)) = V(F_{\beta,\gamma}(\boxplus_\alpha A))$, where $\beta + \gamma \leq 1$,

(c) $V(\boxplus_\alpha G_{\beta,\gamma}(A)) \leq V(G_{\beta,\gamma}(\boxplus_\alpha A))$,

(d) $V(\boxplus_\alpha H_{\beta,\gamma}(A)) = V(H_{\beta,\gamma}(\boxplus_\alpha A))$,

(e) $V(\boxplus_\alpha H^*_{\beta,\gamma}(A)) = V(H^*_{\beta,\gamma}(\boxplus_\alpha A))$,

(f) $V(\boxplus_\alpha J_{\beta,\gamma}(A)) = V(J_{\beta,\gamma}(\boxplus_\alpha A))$,

(g) $V(\boxplus_\alpha J^*_{\beta,\gamma}(A)) = V(J^*_{\beta,\gamma}(\boxplus_\alpha A))$,

(h) $V(\boxtimes_\alpha D_\beta(A)) = V(D_\beta(\boxtimes_\alpha A))$,

(i) $V(\boxtimes_\alpha F_{\beta,\gamma}(A)) = V(F_{\beta,\gamma}(\boxtimes_\alpha A))$, where $\beta + \gamma \leq 1$,

(j) $V(\boxtimes_\alpha G_{\beta,\gamma}(A)) \leq V(G_{\beta,\gamma}(\boxtimes_\alpha A))$,

(k) $V(\boxtimes_\alpha H_{\beta,\gamma}(A)) = V(H_{\beta,\gamma}(\boxtimes_\alpha A))$,

(l) $V(\boxtimes_\alpha H^*_{\beta,\gamma}(A)) = V(H^*_{\beta,\gamma}(\boxtimes_\alpha A))$,

(m) $V(\boxtimes_\alpha J_{\beta,\gamma}(A)) = V(J_{\beta,\gamma}(\boxtimes_\alpha A))$,

(n) $V(\boxtimes_\alpha J^*_{\beta,\gamma}(A)) = V(J^*_{\beta,\gamma}(\boxtimes_\alpha A))$.

The second extension was introduced in [17] by K. Dencheva. She extended the last two operators to the forms:

$$V(\boxplus_{\alpha,\beta} A) = \langle \alpha\mu, \alpha\nu + \beta \rangle, \tag{3.3.5}$$

$$V(\boxtimes_{\alpha,\beta} A) = \langle \alpha\mu + \beta, \alpha\nu \rangle, \tag{3.3.6}$$

where $\alpha, \beta, \alpha + \beta \in [0, 1]$.

Obviously, for every formula A,

$$V(\boxplus A) = V(\boxplus A_{0.5,0.5}),$$

$$V(\boxtimes A) = V(\boxtimes A_{0.5,0.5}),$$

$$V(\boxplus A_\alpha) = V(\boxplus A_{\alpha,1-\alpha}),$$

$$V(\boxtimes A_\alpha) = V(\boxtimes A_{\alpha,1-\alpha}).$$

For every formula A, and for every $\alpha, \beta, \alpha + \beta \in [0, 1]$,

(a) $V(\neg \boxplus_{\alpha,\beta} \neg A) = V(\boxtimes_{\alpha,\beta} A)$,

(b) $V(\neg \boxtimes_{\alpha,\beta} \neg A) = V(\boxplus_{\alpha,\beta} A)$.

For every formula A, and for every $\alpha, \beta \in [0, 1]$, each of the inequalities

(a) $V(\boxplus_{\alpha,\beta} \boxplus_{\alpha,\beta} A) \leq V(\boxplus_{\alpha,\beta} A)$,

(b) $V(\boxtimes_{\alpha,\beta} \boxtimes_{\alpha,\beta} A) \geq V(\boxtimes_{\alpha,\beta} A)$,

is valid if and only if $\alpha + \beta = 1$.

For every formula A, and for every $\alpha \in [0, 1]$,

$$V(\boxplus_{\alpha,\beta} \boxtimes_{\alpha,\beta} A) = V(\boxtimes_{\alpha,\beta} \boxplus_{\alpha,\beta} A) \text{ iff } \beta = 0.$$

For every formula A, and for every $\alpha, \beta, \gamma, \delta \in [0, 1]$ such that $\alpha + \beta, \gamma + \delta \in [0, 1]$,

$$V(\boxplus_{\alpha,\beta} \boxtimes_{\gamma,\delta} A) \leq V(\boxtimes_{\gamma,\delta} \boxplus_{\alpha,\beta} A).$$

Now, the third extension of the above operators is as follows:

$$V(\boxplus_{\alpha,\beta,\gamma} A) = \langle \alpha\mu, \beta\nu + \gamma \rangle, \qquad (3.3.7)$$

$$V(\boxtimes_{\alpha,\beta,\gamma} A) = \langle \alpha\mu + \gamma, \beta\nu \rangle, \qquad (3.3.8)$$

where $\alpha, \beta, \gamma \in [0, 1]$ and $\max(\alpha, \beta) + \gamma \leq 1$.

Obviously, for every formula A,

$$V(\boxplus A) = V(\boxplus A_{0.5,0.5,0.5}),$$

$$V(\boxtimes A) = V(\boxtimes A_{0.5,0.5,0.5}),$$

$$V(\boxplus A_\alpha) = V(\boxplus A_{\alpha,\alpha,1-\alpha}),$$

$$V(\boxtimes A_\alpha) = V(\boxtimes A_{\alpha,1-\alpha}),$$

$$V(\boxplus A_{\alpha,\beta}) = V(\boxplus A_{\alpha,\alpha,\beta}),$$

$$V(\boxtimes A_{\alpha,\beta}) = V(\boxtimes A_{\alpha,\alpha,\beta}).$$

For every formula A, and for every $\alpha, \beta, \gamma \in [0, 1]$ for which $\max(\alpha, \beta) + \gamma \leq 1$,

(a) $V(\neg \boxplus_{\alpha,\beta,\gamma} \neg A)) = V(\boxtimes_{\beta,\alpha,\gamma} A)$,

(b) $V(\neg \boxtimes_{\alpha,\beta,\gamma} \neg A)) = V(\boxplus_{\beta,\alpha,\gamma} A)$.

For every formula A, and for every $\alpha, \beta, \gamma \in [0, 1]$ for which $\max(\alpha, \beta) + \gamma \leq 1$,

(a) $V(\boxplus_{\alpha,\beta,\gamma} \boxplus_{\alpha,\beta,\gamma} A) \leq V(\boxplus_{\alpha,\beta,\gamma} A)$ is valid iff $\beta + \gamma = 1$,

(b) $V(\boxtimes_{\alpha,\beta,\gamma} \boxtimes_{\alpha,\beta,\gamma} A) \geq V(\boxtimes_{\alpha,\beta,\gamma} A)$ is valid iff $\alpha + \gamma = 1$.

For every formula A, and for every $\alpha, \beta, \alpha + \beta \in [0, 1]$,

$$V(\boxplus_{\alpha,\beta,\gamma} \boxtimes_{\alpha,\beta,\gamma} A) = V(\boxtimes_{\alpha,\beta,\gamma} \boxplus_{\alpha,\beta,\gamma} A) \text{ iff } \gamma = 0.$$

For every formula A, and for every $\alpha, \beta, \gamma \in [0, 1]$ for which $\max(\alpha, \beta) + \gamma \leq 1$, the four properties

(a) $V(\boxplus_{\alpha,\beta,\gamma} \square A) = V(\square \boxplus_{\alpha,\beta,\gamma} A)$,

(b) $V(\boxtimes_{\alpha,\beta,\gamma} \square A) = V(\square \boxtimes_{\alpha,\beta,\gamma} A)$,

(c) $V(\boxplus_{\alpha,\beta,\gamma} \diamondsuit A) = V(\diamondsuit \boxplus_{\alpha,\beta,\gamma} A)$,

(d) $V(\boxtimes_{\alpha,\beta,\gamma} \diamondsuit A) = V(\diamondsuit \boxtimes_{\alpha,\beta,\gamma} A)$,

are valid iff $\alpha = \beta$ and $\alpha + \gamma = 1$.

For every two formulas A and B, and for every $\alpha, \beta, \gamma \in [0, 1]$ for which $\max(\alpha, \beta) + \gamma \leq 1$,

(a) $V(\boxplus_{\alpha,\beta,\gamma}(A \wedge B)) = V(\boxplus_{\alpha,\beta,\gamma} A \wedge \boxplus_{\alpha,\beta,\gamma} B)$,

(b) $V(\boxtimes_{\alpha,\beta,\gamma}(A \wedge B)) = V(\boxtimes_{\alpha,\beta,\gamma} A \wedge \boxtimes_{\alpha,\beta,\gamma} B)$,

(c) $V(\boxplus_{\alpha,\beta,\gamma}(A \vee B)) = V(\boxplus_{\alpha,\beta,\gamma} A \vee \boxplus_{\alpha,\beta,\gamma} B)$,

(d) $V(\boxtimes_{\alpha,\beta,\gamma}(A \vee B)) = V(\boxtimes_{\alpha,\beta,\gamma} A \vee \boxtimes_{\alpha,\beta,\gamma} B)$,

For every predicate A, and for every $\alpha, \beta, \gamma \in [0, 1]$ for which $\max(\alpha, \beta) + \gamma \leq 1$,

(a) $V(\boxplus_{\alpha,\beta,\gamma} \exists x A) = V(\exists x \boxplus_{\alpha,\beta,\gamma} A)$,

(b) $V(\boxtimes_{\alpha,\beta,\gamma} \exists x A) = V(\exists x \boxtimes_{\alpha,\beta,\gamma} A)$,

(c) $V(\boxplus_{\alpha,\beta,\gamma} \forall x A) = V(\forall x \boxplus_{\alpha,\beta,\gamma} A)$,

(d) $V(\boxtimes_{\alpha,\beta,\gamma} \forall x A) = V(\forall x \boxtimes_{\alpha,\beta,\gamma} A)$.

A natural extension of the last two operators is the operator

$$V(\boxdot_{\alpha,\beta,\gamma,\delta} A) = \langle \alpha\mu + \gamma, \beta\nu + \delta \rangle, \tag{3.3.9}$$

where $\alpha, \beta, \gamma, \delta \in [0, 1]$ and $\max(\alpha, \beta) + \gamma + \delta \leq 1$.

It is the fourth type of operator from the currently discussed group.

Obviously, for every formula A,

$$V(\boxplus A) = V(\boxdot A_{0.5,0.5,0,0.5}),$$

$$V(\boxtimes A) = V(\boxdot A_{0.5,0.5,0.5,0}),$$

$$V(\boxplus A_\alpha) = V(\boxed{\bullet} A_{\alpha,\alpha,0,1-\alpha}),$$

$$V(\boxtimes A_\alpha) = V(\boxed{\bullet} A_{\alpha,\alpha,1-\alpha,0}),$$

$$V(\boxplus A_{\alpha,\beta}) = V(\boxed{\bullet} A_{\alpha,\alpha,0,\beta}),$$

$$V(\boxtimes A_{\alpha,\beta}) = V(\boxed{\bullet} A_{\alpha,\alpha,\beta,0}),$$

$$V(\boxplus A_{\alpha,\beta,\gamma}) = V(\boxed{\bullet} A_{\alpha,\beta,0,\gamma}),$$

$$V(\boxtimes A_{\alpha,\beta,\gamma}) = V(\boxed{\bullet} A_{\alpha,\beta,\gamma,0}).$$

For every formula A, and for every $\alpha, \beta, \gamma, \delta \in [0, 1]$ for which $\max(\alpha, \beta) + \gamma + \delta \leq 1$:

(a) $V(\neg \boxed{\bullet}_{\alpha,\beta,\gamma,\delta} \neg A) = V(\boxed{\bullet}_{\beta,\alpha,\delta,\gamma} A),$
(b) $V(\boxed{\bullet}_{\alpha,\beta,\gamma,\delta}(\boxed{\bullet}_{\varepsilon,\zeta,\eta,\theta} A)) = V(\boxed{\bullet}_{\alpha\varepsilon,\beta\zeta,\alpha\eta+\gamma,\beta\theta+\delta} A),$
(c) $V(\boxed{\bullet}_{\alpha,\beta,\gamma,\delta} \square A) \geq V(\square \boxed{\bullet}_{\alpha,\beta,\gamma,\delta} A),$
(d) $V(\boxed{\bullet}_{\alpha,\beta,\gamma,\delta} \lozenge A) \leq V(\lozenge \boxed{\bullet}_{\alpha,\beta,\gamma,\delta} A).$

For every pair of formulas A and B, and for every $\alpha, \beta, \gamma, \delta \in [0, 1]$ for which $\max(\alpha, \beta) + \gamma + \delta \leq 1$,

(a) $V(\boxed{\bullet}_{\alpha,\beta,\gamma,\delta}(A \wedge B)) = V(\boxed{\bullet}_{\alpha,\beta,\gamma,\delta} A \wedge \boxed{\bullet}_{\alpha,\beta,\gamma,\delta} B),$
(b) $V(\boxed{\bullet}_{\alpha,\beta,\gamma,\delta}(A \vee B)) = V(\boxed{\bullet}_{\alpha,\beta,\gamma,\delta} A \vee \boxed{\bullet}_{\alpha,\beta,\gamma,\delta} B).$

In [18], G. Çuvalcioğlu introduced the operator $E_{\alpha,\beta}$ by

$$V(E_{\alpha,\beta}(A)) = \langle \beta(\alpha\mu + 1 - \alpha), \alpha(\beta\nu + 1 - \beta) \rangle, \qquad (3.3.10)$$

where $\alpha, \beta \in [0, 1]$, and he studied some of its properties. Obviously,

$$V(E_{\alpha,\beta}(A)) = V(\boxed{\bullet}_{\alpha\beta,\alpha\beta,(1-\alpha)\beta,(1-\beta)\alpha} A).$$

For every predicate A, and for every $\alpha, \beta, \gamma, \delta \in [0, 1]$ for which $\max(\alpha, \beta) + \gamma + \delta \leq 1$,

(a) $V(\boxed{\bullet}_{\alpha,\beta,\gamma,\delta} \exists x A) = V(\exists x \boxed{\bullet}_{\alpha,\beta,\gamma,\delta} A),$
(b) $V(\boxed{\bullet}_{\alpha,\beta,\gamma,\delta} \forall x A) = V(\forall x \boxed{\bullet}_{\alpha,\beta,\gamma,\delta} A).$

A new (potentially final?) extension of the above operators is the operator

$$V(\boxed{\circ}_{\alpha,\beta,\gamma,\delta,\varepsilon,\zeta} A) = \langle \alpha\mu - \varepsilon\nu + \gamma, \beta\nu - \zeta\mu + \delta \rangle, \qquad (3.3.11)$$

where $\alpha, \beta, \gamma, \delta, \varepsilon, \zeta \in [0, 1]$ and

$$\max(\alpha - \zeta, \beta - \varepsilon) + \gamma + \delta \leq 1, \tag{3.3.12}$$

$$\min(\alpha - \zeta, \beta - \varepsilon) + \gamma + \delta \geq 0. \tag{3.3.13}$$

Assume that in the particular cases when $\alpha - \zeta > -\varepsilon, \beta = \delta = 0$ and $\delta - \zeta < \beta, \gamma = \varepsilon = 0$, the inequalities

$$\gamma \geq \varepsilon \quad \text{and} \quad \beta + \delta \leq 1$$

hold. Obviously, for every IFS A,

$$V(\boxplus A) = V(\boxdot_{0.5,0.5,0,0.5,0,0} A),$$

$$V(\boxtimes A) = V(\boxdot_{0.5,0.5,0.5,0,0,0} A),$$

$$V(\boxplus_\alpha A) = V(\boxdot_{\alpha,\alpha,0,1-\alpha,0,0} A),$$

$$V(\boxtimes_\alpha A) = V(\boxdot_{\alpha,\alpha,1-\alpha,0,0,0} A),$$

$$V(\boxplus_{\alpha,\beta} A) = V(\boxdot_{\alpha,\alpha,0,\beta,0,0} A),$$

$$V(\boxtimes_{\alpha,\beta} A) = V(\boxdot_{\alpha,\alpha,\beta,0,0,0} A),$$

$$V(\boxplus_{\alpha,\beta,\gamma} A) = V(\boxdot_{\alpha,\beta,0,\gamma,0,0} A),$$

$$V(\boxtimes_{\alpha,\beta,\gamma} A) = V(\boxdot_{\alpha,\beta,\gamma,0,0,0} A),$$

$$V(\boxdotfilled_{\alpha,\beta,\gamma,\delta} A) = V(\boxdot_{\alpha,\beta,\gamma,\delta,0,0} A),$$

$$V(E_{\alpha,\beta} A) = V(\boxdot_{\alpha\beta,\alpha\beta,\beta(1-\alpha),\alpha(1-\beta)} A).$$

For every formula A, and for every $\alpha, \beta, \gamma, \delta, \varepsilon, \zeta \in [0, 1]$ for which (3.3.12) and (3.3.13) are valid, the equality

$$V(\neg\boxdot_{\alpha,\beta,\gamma,\delta,\varepsilon,\zeta}\neg A) = V(\boxdot_{\beta,\alpha,\delta,\gamma,\zeta,\varepsilon} A)$$

holds.

For every formula A, and for every $\alpha_1, \beta_1, \gamma_1, \delta_1, \varepsilon_1, \zeta_1, \alpha_2, \beta_2, \gamma_2, \delta_2, \varepsilon_2, \zeta_2 \in$ [0, 1] for which conditions that are similar to (3.3.12) and (3.3.13) are valid, the equality

$$V(\boxed{\circ}_{\alpha_1,\beta_1,\gamma_1,\delta_1,\varepsilon_1,\zeta_1}(\boxed{\circ}_{\alpha_2,\beta_2,\gamma_2,\delta_2,\varepsilon_2,\zeta_2} A))$$

$$=V(\boxed{\circ}_{\alpha_1\alpha_2+\varepsilon_1\zeta_2,\beta_1\beta_2+\zeta_1\varepsilon_2,\alpha_1\gamma_2-\varepsilon_1\delta_2+\gamma_1,\beta_1\delta_2-\zeta_1\gamma_2+\delta_1,\alpha_1\varepsilon_2+\varepsilon_1\beta_2,\beta_1\zeta_2+\zeta_1\alpha_2} A)$$

holds.

It must be noted that the equalities

$$V(\boxed{\circ}_{\alpha,\beta,\gamma,\delta,\varepsilon,\zeta}(A \wedge B)) = V(\boxed{\circ}_{\alpha,\beta,\gamma,\delta,\varepsilon,\zeta} A \wedge \boxed{\circ}_{\alpha,\beta,\gamma,\delta,\varepsilon,\zeta} B)$$

and

$$V(\boxed{\circ}_{\alpha,\beta,\gamma,\delta,\varepsilon,\zeta}(A \vee B)) = V(\boxed{\circ}_{\alpha,\beta,\gamma,\delta,\varepsilon,\zeta} A \vee \boxed{\circ}_{\alpha,\beta,\gamma,\delta,\varepsilon,\zeta} B),$$

which are valid for operator $\boxed{\bullet}_{\alpha,\beta,\gamma,\delta}$, are not always valid for $\boxed{\circ}_{\alpha,\beta,\gamma,\delta,\varepsilon,\zeta}$.

Open Problem 18 Check the validity of the above formulas for the case of all intuitionistic fuzzy conjunctions, disjunctions, implications and negations.

Following [9], we formulate and prove the following:

Theorem 3.3.1 *Operators* $X_{a,b,c,d,e,f}$ *and* $\boxed{\circ}_{\alpha,\beta,\gamma,\delta,\varepsilon,\zeta}$ *are equivalent.*

Proof Let $a, b, c, d, e, f \in [0, 1]$ and satisfy (6.24) and (6.25). Let

$$\alpha = a - b, \quad \beta = d - e, \quad \gamma = b, \quad \delta = e, \quad \varepsilon = bc, \quad \zeta = ef.$$

Also, let

$$X \equiv \alpha\mu - \varepsilon\nu + \gamma = (a - b)\mu - bc\nu + b,$$

$$Y \equiv \beta\nu - \zeta\mu + \delta = (d - e)\nu - ef\mu + e.$$

Then,

$$X \geq (a - b).0 - bc.1 + b = b(1 - c) \geq 0,$$

$$X \leq (a - b).1 - bc.0 + b = a \leq 1,$$

$$Y \geq (d - e).0 - ef.1 + e = e(1 - f) \geq 0,$$

$$Y \leq (d - e).1 - ef.0 + e = d \leq 1$$

and

$$X + Y = (a - b)\mu - bc\nu + b + (d - e)\nu - ef\mu + e$$

$$= (a - b - ef)\mu + (d - e - bc)\nu + b + e$$

$$\leq (a - b - ef)\mu + (d - e - bc)(1 - \mu) + b + e$$

$$= d - e - bc + b + e + (a - b - ef - d + e + bc)\mu$$

$$\leq d - bc + b + a - b - ef - d + e + bc$$

$$= a - ef + e = \alpha + \gamma - \zeta + \delta \leq 1$$

(from (3.3.12)).

Thus, we obtain

$$V(\boxed{\circ}_{\alpha,\beta,\gamma,\delta,\varepsilon,\zeta} A) = \langle \alpha\mu - \varepsilon\nu + \gamma, \beta\nu - \zeta\mu + \delta \rangle$$

$$= \langle x, (a - b)\mu - bc\nu + b, (d - e)\nu - ef\mu + e \rangle$$

$$= \langle x, a\mu + b(1 - \mu - c\nu),$$

$$d\nu + e(1 - f\mu - \nu) \rangle$$

$$= V(X_{a,b,c,d,e,f}(A)).$$

Conversely, let $\alpha, \beta, \gamma, \delta, \varepsilon, \zeta \in [0, 1]$ and satisfy (3.3.12) and (3.3.13). From (3.3.13) it follows that for $\alpha = \beta = \delta = \zeta = 0 : \varepsilon \leq \gamma$, while for $\alpha = \beta = \gamma = \varepsilon = 0 : \zeta \leq \delta$; from (3.3.12) it follows that for $\beta = \delta = \varepsilon = \zeta = 0 : \alpha + \gamma \leq 1$, while for $\alpha = \gamma = \varepsilon = \zeta = 0 : \beta + \delta \leq 1$. Then, let

$$a = \alpha + \gamma \ (\leq 1),$$

$$b = \gamma,$$

$$c = \frac{\varepsilon}{\gamma} \ (\leq 1),$$

$$d = \beta + \delta \ (\leq 1),$$

$$e = \delta,$$

$$f = \frac{\zeta}{\delta} \ (\leq 1).$$

Let

$$X \equiv a\mu + b(1 - \mu - c\nu) = (\alpha + \gamma)\mu + \gamma\left(1 - \mu - \frac{\varepsilon}{\gamma}\nu\right),$$

$$Y \equiv d\nu + e(1 - f\mu - \nu) = (\beta + \delta)\nu + \delta\left(1 - \frac{\zeta}{\delta}\mu - \nu\right).$$

Then, we obtain,

$$0 \le \gamma - \varepsilon \le X = \alpha\mu + \gamma - \varepsilon\nu \le \alpha + \gamma \le 1,$$

$$0 \le \delta - \zeta \le Y = \beta\nu + \delta - \zeta\mu \le \beta + \delta \le 1,$$

$$X + Y = \alpha\mu + \gamma - \varepsilon\nu + \beta\nu + \delta - \zeta\mu$$

$$= (\alpha - \zeta)\mu - (\beta - \varepsilon)\nu + \gamma + \delta$$

$$\le (\alpha - \zeta)\mu - (\beta - \varepsilon)(1 - \mu) + \gamma + \delta$$

$$= (\alpha - \zeta + \beta - \varepsilon)\mu - \beta + \gamma + \delta + \varepsilon$$

$$\le \alpha - \zeta + \beta - \varepsilon - \beta + \gamma + \delta + \varepsilon$$

$$= \alpha - \zeta + \gamma + \delta$$

$$\le \max(\alpha - \zeta, \beta - \varepsilon) + \gamma + \delta \le 1$$

(from (3.3.12)).

Then, we obtain

$$V(X_{a,b,c,d,e,f}(A))$$

$$= \langle a\mu + b(1 - \mu - c\nu), d\nu + e(1 - f\mu - \nu)\rangle$$

$$= \langle(\alpha + \gamma)\mu + \gamma(1 - \mu - \frac{\varepsilon}{\gamma}\nu), (\beta + \delta)\nu + \delta(1 - \frac{\zeta}{\delta}\mu - \nu)\rangle$$

$$= \langle(\alpha + \gamma)\mu + \gamma - \gamma\mu - \varepsilon\nu, (\beta + \delta)\nu + \delta - \zeta\mu - \delta\nu\rangle$$

$$= \langle\alpha\mu - \varepsilon\nu + \gamma, \beta\nu - \zeta\mu + \delta\rangle$$

$$= V(\boxed{\circ}_{\alpha,\beta,\gamma,\delta,\varepsilon,\zeta}A).$$

Therefore, the two operators are equivalent.

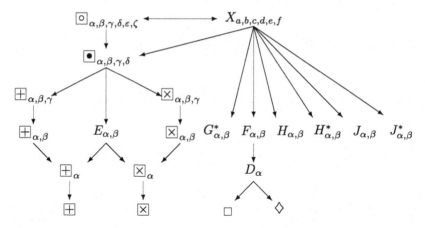

Fig. 3.15 Relations among modal operators

Finally, we construct Fig. 3.15 in which

$$\text{``}X \longrightarrow Y\text{''}$$

denotes that operator X represents operator Y, while the converse is not valid, and

$$\text{``}X \longleftrightarrow Y\text{''}$$

denotes that each of the operators represents the other.

Following [19], we introduce the following new operator from modal type, that is a modification of the above discussed operators. It has the form

$$V(\otimes_{\alpha,\beta,\gamma,\delta}A) = \langle \alpha a + \gamma b, \beta a + \delta b \rangle,$$

where $\alpha, \beta, \gamma, \delta \in [0, 1]$ and $\alpha + \beta \leq 1, \gamma + \delta \leq 1$.

First, we check that the new operator generates an intuitionistic fuzzy pair. Indeed,

$$0 \leq \alpha a + \gamma b \leq a + b \leq 1,$$

$$0 \leq \beta a + \delta b \leq a + b \leq 1$$

and

$$0 \leq \alpha a + \gamma b + \beta a + \delta b$$

$$= (\alpha + \beta)a + (\gamma + \delta)b$$

$$\leq a + b \leq 1.$$

Second, it is easy to see that

$$\otimes_{1,0,0,1} A = A,$$

$$\otimes_{0,1,1,0} A = \neg_1 A.$$

Theorem 3.3.2 *For every formula A, for every four real numbers $\alpha, \beta, \gamma, \delta \in [0, 1]$ such that $\alpha + \beta \leq 1, \gamma + \delta \leq 1$ and for negation $\neg_1$,*

$$V(\neg_1 \otimes_{\alpha,\beta,\gamma,\delta} \neg_1 A) = V(\otimes_{\delta,\gamma,\beta,\alpha} A).$$

Proof We obtain sequentially that

$$V(\neg_1 \otimes_{\alpha,\beta,\gamma,\delta} \neg_1 A)$$

$$= \neg_1 \otimes_{\alpha,\beta,\gamma,\delta} \langle b, a \rangle$$

$$= \neg_1 \langle \alpha b + \gamma a, \beta b + \delta a \rangle$$

$$= \langle \beta b + \delta a, \alpha b + \gamma a \rangle$$

$$= \otimes_{\delta,\gamma,\beta,\alpha} A.$$

This completes the proof.   □

Theorem 3.3.3 *For every two formulas A and B an for every four real numbers $\alpha, \beta, \gamma, \delta \in [0, 1]$ such that $\alpha + \beta \leq 1, \gamma + \delta \leq 1$,*

(a) $V(\otimes_{\alpha,\beta,\gamma,\delta}(A \vee B)) = V(\otimes_{\alpha,\beta,\gamma,\delta} A \vee \otimes_{\alpha,\beta,\gamma,\delta} B)$,
(b) $V(\otimes_{\alpha,\beta,\gamma,\delta}(A \wedge B)) = V(\otimes_{\alpha,\beta,\gamma,\delta} A \wedge \otimes_{\alpha,\beta,\gamma,\delta} B)$.

Proof For (a), first, we obtain that

$$V(\otimes_{\alpha,\beta,\gamma,\delta}(A \vee B)) = \otimes_{\alpha,\beta,\gamma,\delta} \langle \max(a, c), \min(b, d) \rangle$$

$$= \langle \alpha \max(a, c) + \gamma \min(b, d), \beta \max(a, c) + \delta \min(b, d) \rangle.$$

Second, we obtain that
$$V(\otimes_{\alpha,\beta,\gamma,\delta} A \vee \otimes_{\alpha,\beta,\gamma,\delta} B)$$

$$= \langle \alpha a + \gamma b, \beta a + \delta b \rangle \vee \langle \alpha c + \gamma d, \beta c + \delta.d \rangle$$

$$= \langle \max(\alpha a + \gamma b, \alpha c + \gamma d), \min(\beta a + \delta b, \beta c + \delta d) \rangle$$

Let

$$X \equiv \max(\alpha a + \gamma b, \alpha c + \gamma d) - \alpha \max(a, c) - \gamma \min(\nu_A(x), d)$$

Now, for a, c, b, d we must study the following four cases.

Case 1. $a \geq c, b \geq d$:

$$X = \max(\alpha a + \gamma b, \alpha c + \gamma d) - \alpha a - \gamma d$$

$$\geq \alpha a + \gamma b - \alpha a - \gamma d \geq 0$$

Case 2. $a \geq c, b < d$:

$$X = \max(\alpha a + \gamma b, \alpha c + \gamma d) - \alpha a - \gamma b \geq 0$$

Case 3. $a < c, b \geq d$:

$$X = \max(\alpha a + \gamma b, \alpha c + \gamma d) - \alpha c - \gamma d \geq 0$$

Case 4. $a < c, b < d$:

$$X = \max(\alpha a + \gamma b, \alpha c + \gamma d) - \alpha c) - \gamma b$$

$$\geq \alpha c + \gamma d - \alpha c) - \gamma b \geq 0$$

Assertion (b) is proved analogously.   □

The proofs of the next assertion follow by analogy.

Theorem 3.3.4 *For every formula A and for every four real numbers* $\alpha, \beta, \gamma, \delta \in [0, 1]$ *such that* $\alpha + \beta \leq 1, \gamma + \delta \leq 1$:

(a) $V(\square \otimes_{\alpha,\beta,\gamma,\delta} A) \leq V(\otimes_{\alpha,\beta,\gamma,\delta} \square A)$,
(b) $V(\otimes_{\alpha,\beta,\gamma,\delta} \lozenge A) \leq V(\lozenge \otimes_{\alpha,\beta,\gamma,\delta} A)$.

Theorem 3.3.5 *Let A be a formula, such that* $V(A) = \langle \mu, \nu \rangle$ *and let* $a, b, c, d, e, f, g, h \in [0, 1]$, *so that* $a + b, c + d, e + f, g + h \in [0, 1]$. *Then,*

$$V(\otimes_{e,f,g,h}(\otimes_{a,b,c,d}(A))) = V(\otimes_{ae+bg,af+bh,ce+dg,cf+dh}(A)). \tag{3.3.14}$$

Proof Let formula A and the real numbers a, b, c, d, e, f, g, h satisfy the conditions for operator $\otimes$. Then

$$V(\otimes_{e,f,g,h}(\otimes_{a,b,c,d}(A)))$$

$$= \otimes_{e,f,g,h}\langle x, a\mu + c\nu, b\mu + d\nu \rangle$$

$$= \langle x, ae\mu + ce\nu + bg\mu + dg\nu, af\mu + cf\nu + bh\mu + dh\nu \rangle$$

$$= \langle x, (ae + bg)\mu + (ce + dg)\nu, (af + bh)\mu + (cf + dh)\nu \rangle$$

$$= V(\otimes_{ae+bg,af+bh,ce+dg,cf+dh}(A)).$$

Therefore, (3.3.14) is valid. □

Theorem 3.3.6 *Let A be a formula, such that* $V(A) = \langle \mu, \nu \rangle$ *and let* $a, d, e, h \in (0, 1], b, c, f, g \in (0, 1],$ *so that* $a + b, c + d, e + f, g + h \in [0, 1]$ *and*

$$bg = cf, \tag{3.3.15}$$

$$ag + ch = ce + dg. \tag{3.3.16}$$

Then,

$$V(\otimes_{e,f,g,h}(\otimes_{a,b,c,d}(A))) = V(\otimes_{a,b,c,d}(\otimes_{e,f,g,h}(A))). \tag{3.3.17}$$

Proof Let formula A and the real numbers a, b, c, d, e, f, g, h satisfy the conditions of the theorem. First, we see, that from (3.3.15) and (3.3.16) it follows:

$$af + bh - be - df = af + \frac{cf}{g}h - \frac{cf}{g}e - df = \frac{f}{g}(ag + ch - ce - dg) = 0.$$

But, by the above conditions, $f, g > 0$. Therefore,

$$af + bh - be - df = 0,$$

i.e.,

$$af + bh = be + df. \tag{3.3.18}$$

Now,

$$V(\otimes_{e,f,g,h}(\otimes_{a,b,c,d}(A))) = \otimes_{e,f,g,h}\langle a.\mu + c.\nu, b.\mu + d.\nu \rangle$$

$$= \langle a.e.\mu + c.e.\nu + b.g.\mu + d.g.\nu, a.f.\mu + c.f.\nu + b.h.\mu + d.h.\nu \rangle$$

(from (3.3.16) and (3.3.18))

$$= \langle a.e.\mu + a.g.\nu + c.f.\mu + c.h.\nu, b.e.\mu + b.g.\nu + d.f.\mu + d.h.\nu \rangle$$

$$= \otimes_{a,b,c,d}\langle e.\mu + g.\nu, f.\mu + h.\nu \rangle$$

$$= V(\otimes_{a,b,c,d} \otimes_{e,f,g,h}(A)).$$

Therefore, (3.3.17) is valid. □

Theorem 3.3.7 *Let A be a formula, $\alpha, \beta, \gamma, \delta \in [0, 1]$, so that $\alpha + \beta, \gamma + \delta \in [0, 1]$, $a, b, c, d, e, f \in [0, 1]$ so that $a + e - e.f \le 1$, $b + d - b.c \le 1$ and $b + e \le 1$. Then,*

(a) $\boxed{\circ}_{a,b,c,d,e,f}(\otimes_{\alpha,\beta,\gamma,\delta}(A)) = \boxed{\circ}_{a\alpha-e\beta,b\beta-f\alpha,c,d,e\delta-a\gamma,f\gamma-b\delta}(A),$

(b) $\otimes_{\alpha,\beta,\gamma,\delta}(\boxed{\circ}_{a,b,c,d,e,f}(A)) = \boxed{\circ}_{a\alpha-f\gamma,a\beta-f\delta,c\alpha+d\gamma,c\beta+d\delta,e\alpha-b\gamma,e\beta-b\delta}(A).$

The following two open problems are interesting:

Open Problem 19 Can operator $\otimes_{\alpha,\beta,\gamma,\delta}$ be represented by the extended modal operators?

Open Problem 20 Can operator $\otimes_{\alpha,\beta,\gamma,\delta}$ be used for representation of some types of modal operators?

3.4 Intuitionistic Fuzzy Level Operators

Following [9], here we introduce the following two intuitionistic fuzzy level operators for each formula A with evaluation $V(A) = \langle a, b \rangle$:

$$V(P_{\alpha,\beta}(A)) = P_{\alpha,\beta}(V(A)) = \langle \max(a, \alpha), \min(b, \beta) \rangle,$$

$$V(Q_{\alpha,\beta}(A)) = Q_{\alpha,\beta}(V(A)) = \langle \min(a, \alpha), \max(b, \beta) \rangle.$$

We must note, that for every formula A

$$V(P_{\alpha,\beta}(A)) = V(A) \vee \langle \alpha, \beta \rangle$$

and

$$V(Q_{\alpha,\beta}(A)) = V(A) \wedge \langle \alpha, \beta \rangle.$$

Theorem 3.4.1 *For every formula A and for every $\alpha, \beta, \gamma, \delta \in [0, 1]$, such that $\alpha + \beta \le 1$, $\gamma + \delta \le 1$:*

(a) $V(\neg P_{\alpha,\beta}(\neg A)) = V(Q_{\beta,\alpha}(A));$

(b) $V(P_{\alpha,\beta}(Q_{\gamma,\delta}(A))) = V(Q_{\max(\alpha,\gamma),\min(\beta,\delta)}(P_{\alpha,\beta}(A)));$

(c) $V(Q_{\alpha,\beta}(P_{\gamma,\delta}(A))) = V(P_{\min(\alpha,\gamma),\max(\beta,\delta)}(Q_{\alpha,\beta}(A)));$

(d) $V(P_{\alpha,\beta}(P_{\gamma,\delta}(A))) = V(P_{\max(\alpha,\gamma),\min(\beta,\delta)}(A));$

(e) $V(Q_{\alpha,\beta}(Q_{\gamma,\delta}(A))) = V(Q_{\min(\alpha,\gamma),\max(\beta,\delta)}(A)).$

Proof (b) Let A be a formula. Then,

$$V(P_{\alpha,\beta}(Q_{\gamma,\delta}(A)))$$
$$= V(P_{\alpha,\beta}(\langle \min(\gamma, a), \max(\delta, b)\rangle))$$
$$= \langle \max(\alpha, \min(\gamma, a)), \min(\beta, \max(\delta, b))\rangle$$
$$= \langle \min(\max(\alpha, \gamma), \max(\alpha, a)), \max(\min(\beta, \delta), \max(\beta, a))\rangle$$
$$= Q_{\max(\alpha,\gamma),\min(\beta,\delta)}(\langle \max(\alpha, a), \max(\beta, b)\rangle))$$
$$= V(Q_{\max(\alpha,\gamma),\min(\beta,\delta)}(P_{\alpha,\beta}(A))).$$

This completes the proof.   ☐

Theorem 3.4.2 *For every two formulas A and B, and for every $\alpha, \beta \in [0, 1]$, such that $\alpha + \beta \leq 1$:*

(a) $V(P_{\alpha,\beta}(A \wedge B)) = V(P_{\alpha,\beta}(A)) \wedge V(P_{\alpha,\beta}(B))$,
(b) $V(P_{\alpha,\beta}(A \vee B)) = V(P_{\alpha,\beta}(A)) \vee V(P_{\alpha,\beta}(B))$,
(c) $V(Q_{\alpha,\beta}(A \wedge B)) = V(Q_{\alpha,\beta}(A)) \wedge V(Q_{\alpha,\beta}(B))$,
(d) $V(Q_{\alpha,\beta}(A \vee B)) = V(Q_{\alpha,\beta}(A)) \vee V(Q_{\alpha,\beta}(B))$.

Proof (a) Let A and B be two formulas. Then,

$$V(P_{\alpha,\beta}(A \wedge B))$$
$$= P_{\alpha,\beta}(\langle \min(a, c), \max(b, d)\rangle)$$
$$= \langle \max(\alpha, \min(a, c)), \min(\beta, \max(b, d))\rangle$$
$$= \langle \min(\max(\alpha, a), \max(\alpha, c)), \max(\min(\beta, b), \min(\beta, d))\rangle$$
$$= V(P_{\alpha,\beta}(A) \wedge P_{\alpha,\beta}(B)).$$

This completes the proof.   ☐

Theorem 3.4.3 *Let A be a formula and x be a variable. Then, for every $\alpha, \beta \in [0, 1]$, such that $\alpha + \beta \leq 1$:*

$$(a)\quad V(\exists x\, P_{\alpha,\beta}(A)) = V(P_{\alpha,\beta}(\exists x A)),$$
$$(b)\quad V(\forall x\, Q_{\alpha,\beta}(A)) = V(Q_{\alpha,\beta}(\forall x A)).$$

Proof (a) Let A be a formula. Then,

$$V(\exists x\, P_{\alpha,\beta}(A))$$
$$= \exists x \langle \max(\alpha, \mu(A)), \min(\beta, \nu(A))\rangle$$
$$= \langle \max_x(\max(\alpha, \mu(A))), \min_x(\min(\beta, \nu(A)))\rangle$$
$$= \langle \max(\alpha, \max_x(\mu(A))), \min(\beta, \min_x(\nu(A)))\rangle$$
$$= V(P_{\alpha,\beta}(\exists x A)).$$

This completes the proof. Assertion (b) is proved by analogy.   ☐

3.5   Pseudo-fixed Points of the Intuitionistic Fuzzy Operators and Quantifiers

Let S be a set of propositions (or more general, formulas) and let $V : S \to [0, 1] \times [0, 1]$, be defined for every $A \in S$ as in Sect. 1.1.

Let for operator Y and for IFP $\langle a, b \rangle$:

$$Y(\langle a, b \rangle) = \langle a, b \rangle.$$

Then, we call that the IFP is a fixed point for operator Y. But, when operator Y is defined over elements of S, i.e., when for formula A

$$V(Y(A)) = \langle \mu(Y(A)), \nu(Y(A)) \rangle,$$

then we will call that A is a pseudo-fixed point for operator Y. In this case, the equality

$$\langle \mu(Y(A)), \nu(Y(A)) \rangle = \langle \mu(A), \nu(A) \rangle \tag{3.5.1}$$

holds (see [20]).

Obviously, if (3.5.1) is valid for IFP $V(A) = \langle a, b \rangle$, then, $\langle a, b \rangle$ is a fixed point for operator Y.

Below, we determine all pseudo-fixed points of all quantifiers and operators, defined in Chaps. 2 and 3.

Theorem 3.5.1 *For all $\alpha, \beta \in [0, 1]$ the pseudo-fixed point(s) of:*

(a) $\exists$ are all elements $A \in S$ for which $V(A) = \langle 1, 0 \rangle$,
(b) $\forall$ are all elements $A \in S$ for which $V(A) = \langle 0, 1 \rangle$,
(c) $\exists_\mu$ are all elements $A \in S$ for which

$$\mu(A) = \sup_{x \in S} \mu(x)$$

and in the more general case, all elements $A \in S$ for which $V(A) = \langle 1, 0 \rangle$,
(d) $\exists_\nu$ are all elements $A \in S$ for which

$$\nu(A) = \inf_{x \in S} \nu(x)$$

and in the more general case, all elements $A \in S$ for which $V(A) \in [0, 1] \times \{0\}$,
(e) $\forall_\mu$ are all elements $A \in S$ for which

$$\mu(A) = \inf_{x \in S} \mu(x)$$

and in the more general case, all elements $A \in S$ for which $V(A) \in \{0\} \times [0, 1]$, *(f)* $\forall_\nu$ *are all elements* $A \in S$ *for which*

$$\nu(A) = \sup_{x \in S} \nu(x)$$

and in the more general case, all elements $A \in S$ *for which* $V(A) = \langle 0, 1 \rangle$,

(g) $\square, \diamond, \bigcirc$ *are all elements* $A \in S$ *for which* $\mu(A) + \nu(A) = 1$,

(h) D_α *are all elements* $A \in S$ *for which* $\mu(A) + \nu(A) = 1$,

(i) $F_{\alpha,\beta}$ *are all elements* $A \in S$ *for which* $\mu(A) + \nu(A) = 1$ *and* $\alpha + \beta \leq 1$,

(j) $G_{\alpha,\beta}$ *are all elements* $A \in S$ *for which* $\mu(A) = \nu(A) = 0$,

(k) $H_{\alpha,\beta}, H^*_{\alpha,\beta}$ *are all elements* $A \in S$ *for which* $\mu(A) = 0$ *and* $\nu(A) = 1$,

(l) $J_{\alpha,\beta}, J^*_{\alpha,\beta}$ *are all elements* $A \in S$ *for which* $\mu(A) = 1$ *and* $\nu(A) = 0$,

(m) $\boxplus, \boxplus_\alpha$ *are all elements* $A \in S$ *for which* $\mu(A) = 0$ *and* $\nu(A) = 1$,

(n) $\boxtimes, \boxtimes_\alpha$ *are all elements* $A \in S$ *for which* $\mu(A) = 1$ *and* $\nu(A) = 0$,

(o) $\boxplus_{\alpha,\beta}$ *are all elements* $A \in S$ *for which* $\mu(A) = 0, \nu(A) = 1$ *and* $\alpha + \beta = 1$,

(p) $\boxtimes_{\alpha,\beta}$ *are all elements* $A \in S$ *for which* $\mu(A) = 1, \nu(A) = 0$ *and* $\alpha + \beta = 1$,

(q) $P_{\alpha,\beta}$ *are all elements* $A \in S$ *for which* $\alpha \leq \mu(A) = 1$ *and* $0 \leq \nu(A) \leq \beta$,

(r) $Q_{\alpha,\beta}$ *are all elements* $A \in S$ *for which* $0 \leq \mu(A) = \alpha$ *and* $\beta \leq \nu(A) \leq 1$.

References

1. Atanassov K. Intuitionistic fuzzy sets, VII ITKR's Session; 1983 (Deposed in Central Sci. - Techn. Library of Bulg. Acad. of Sci., 1697/84) (in Bulg.), Reprinted: Int J Bioautomation. 2016;2(S1):S1–S6.
2. Blackburn P, van Bentham J, Wolter F. Modal logic. Amsterdam: North Holland; 2006.
3. Feys R. Modal logics. Paris: Gauthier-Villars; 1965.
4. Fitting M, Mendelsohn R. First order modal logic. Dordrecht: Kluwer; 1998.
5. Mints G. A short introduction to modal logic. Chicago: University of Chicago Press; 1992.
6. Atanassov K. Elements of intuitionistic fuzzy logics. Part II: Intuitionistic fuzzy modal logics. Adv Stud Contemp Math. 2002;5(1):1–13.
7. Atanassov K. Intuitionistic fuzzy sets. Heidelberg: Springer; 1999.
8. Atanassov K. A new topological operator over intuitionistic fuzzy sets. Notes on Intuitionistic Fuzzy Sets. 2015;21(3):90–2.
9. Atanassov K. On intuitionistic fuzzy sets theory. Berlin: Springer; 2012.
10. Atanassov K. A short remark on intuitionistic fuzzy operators $X_{a,b,c,d,e,f}$ and $x_{a,b,c,d,e,f}$. Notes on Intuitionistic Fuzzy Sets. 2013;19(1):54–6.
11. Atanassov K. A property of the intuitionistic fuzzy modal logic operator $X_{a,b,c,d,e,f}$. Notes on Intuitionistic Fuzzy Sets. 2015;21(1):1–5.
12. Atanassov K. Some operators on intuitionistic fuzzy sets. In: Kacprzyk J, Atanassov K, editors. Proceedings of the First International Conference on Intuitionistic Fuzzy Sets. Sofia, Oct 18-19, 1997; Notes on Intuitionistic Fuzzy Sets. 1997;3(4):28–33. http://ifigenia.org/wiki/issue:nifs/3/4/28-33.
13. Atanassov K. On one type of intuitionistic fuzzy modal operators. Notes on Intuitionistic Fuzzy Sets. 2005;11(5):24–28. http://ifigenia.org/wiki/issue:nifs/11/5/24-28.
14. Atanassov K. The most general form of one type of intuitionistic fuzzy modal operators. Notes on Intuitionistic Fuzzy Sets. 2006;12(2):36–38. http://ifigenia.org/wiki/issue:nifs/12/2/36-38.

15. Atanassov K. Some properties of the operators from one type of intuitionistic fuzzy modal operators. Adv Stud Contemp Math. 2007;15(1):13–20.
16. Atanassov K. The most general form of one type of intuitionistic fuzzy modal operators. Part 2. Notes on Intuitionistic Fuzzy Sets. 2008;14(1):27–32. http://ifigenia.org/wiki/issue:nifs/14/1/27-32.
17. Dencheva K. Extension of intuitionistic fuzzy modal operators $\boxtimes$ and $\boxtimes$. Proceedings of the Second Int. IEEE Symposium: Intelligent Systems, Varna, June 22–24, vol. 3; 2004. p. 21–22.
18. Çuvalcioğlu G. Some properties of $E_{\alpha,\beta}$ operator. Adv Stud Contemp Math. 2007;14(2):305–310.
19. Atanassov K, Çuvalcioğlu G, Atanassova V. A new modal operator over intuitionistic fuzzy sets. Notes on Intuitionistic Fuzzy Sets. 2014;20(5):1–8.
20. Atanassov, K. On Pseudo-fixed Points of the Intuitionistic Fuzzy Quantifiers and Operators, Proceedings of the 8th European Symposium on Computational Intelligence and Mathematics, Sofia, Bulgaria, 5–8 October 2016:66–76.

Chapter 4
Temporal and Multidimensional Intuitionistic Fuzzy Logics

The first results in temporal intuitionistic fuzzy logic appeared in 1990 (see [1]) on the basis of ideas from [2]. However, the first example for their application was only proposed as early as 15 years later, in [3]. The concept of the temporal IFL was extended to the concept of multidimensional intuitionistic fuzzy logic in a series of papers of the author together with E. Szmidt and J. Kacprzyk.

4.1 Temporal Intuitionistic Fuzzy Logic

Let A be a formula and V be a truth-value function, which maps to A the ordered pair

$$V(A, t) = \langle \mu(A, t), \nu(A, t) \rangle,$$

where $\mu(A, t), \nu(A, t) \in [0, 1]$,

$$\mu(A, t) + \nu(A, t) \leq 1,$$

and $t \in T$ is a fixed time-moment, where T is a fixed set which we shall call "time-scale" and it is strictly ordered by the relation "$<$".

Let

$$T'(t) = \{t' \mid t' \in T \,\&\, t' < t\},$$

$$T''(t) = \{t'' \mid t'' \in T \,\&\, t'' > t\}.$$

In [1], for a given formula A and a time-moment t, the author defined the temporal intuitionistic fuzzy operators P, F, H, G, which are analogues of the operators from [2], and for which:

- $V(P(A, t)) = P(V(A), t) = \langle \mu(A, t'), \nu(A, t') \rangle,$

© Springer International Publishing AG 2017
K.T. Atanassov, *Intuitionistic Fuzzy Logics*, Studies in Fuzziness
and Soft Computing 351, DOI 10.1007/978-3-319-48953-7_4

where $t' \in T'$ satisfies the conditions:

(a) $\mu(A, t') - \nu(A, t') = \max_{t^* \in T'}(\mu(A, t^*) - \nu(A, t^*))$,

(b) if there exists more than one such element of T', then, t' is maximal;

• $V(F(A, t)) = \langle \mu(A, t''), \nu(A, t'') \rangle$,
  where $t'' \in T''$ satisfies the conditions:

(a) $\mu(A, t'') - \nu(A, t'') = \max_{t^* \in T''}(\mu(A, t^*) - \nu(A, t^*))$,

(b) if there exists more than one such element of T'', then, t'' is minimal;

• $V(H(A, t)) = \langle \mu(A, t'), \nu(A, t') \rangle$,
  where $t' \in T'$ satisfies the conditions:

(a) $\mu(A, t') - \nu(A, t') = \min_{t^* \in T'}(\mu(A, t^*) - \nu(A, t^*))$,

(b) if there exists more than one such element of T', then, t' is maximal;

• $V(G(A, t)) = \langle \mu(A, t''), \nu(A, t'') \rangle$,
  where $t'' \in T''$ satisfies the conditions:

(a) $\mu(A, t'') - \nu(A, t'') = \min_{t^* \in T''}(\mu(A, t^*) - \nu(A, t^*))$,

(b) if there exists more than one such element of T'', then, t'' is minimal.

In each of these four definitions, if time intervals T' or T'' are infinite, operations "max" and "min" must be changed with operations "sup" and "inf", respectively.

Theorem 4.1.1 *For every formula A and for every time-moment t:*

(a) $V(H(A, t)) = V(\neg(P(\neg A), t)))$;
(b) $V(G(A, t)) = V(\neg(F(\neg A), t)))$.

Proof (a) Let the formula A and the time-moment t be given. Then,

$$V(\neg(F(\neg A), t))) = \langle \mu(A, t'), \nu(A, t') \rangle$$

where t' is the maximal element of T' for which:

$$\nu(A, t') - \mu(A, t') = \max_{t* \in T'}(\nu(A, t*) - \mu(A, t*)).$$

Therefore, t' is the maximal element of T' for which:

$$\mu(A, t') - \nu(A, t') = \min_{t* \in T'}(\mu(A, t*) - \nu(A, t*)),$$

i.e.,

$$\langle \mu(A, t'), \nu(A, t') \rangle = V(H(A, t)).$$

This completes the proof. Assertion (b) is proved similarly. □

Theorem 4.1.2 *For every two formulas A and B, for every time-moment t for implication $\rightarrow_4$:*

(a) $H(A \rightarrow_4 B, t) \rightarrow_4 (P(A, t) \rightarrow_4 P(B, t))$;
(b) $G(A \rightarrow_4 B, t) \rightarrow_4 (F(A, t) \rightarrow_4 F(B, t))$;
(c) $\neg(P(\neg(A \rightarrow_4 B), t)) \rightarrow_4 (P(A, t) \rightarrow_4 P(B, t))$;
(d) $\neg(F(\neg(A \rightarrow_4 B), t)) \rightarrow_4 (F(A, t) \rightarrow_4 F(B, t))$.

are IFTs.

Proof (a) Let the formulas A and B, and the time-moment t be given. Then,

$$H(A \rightarrow_4 B, t) \rightarrow_4 (P(A, t) \rightarrow_4 P(B, t))$$

$$= H(\langle \max(\nu(A), \mu(B)), \min(\mu(A), \nu(B))\rangle, t)$$

$$\rightarrow_4 (\langle \mu(A, t_1), \nu(A, t_1)\rangle \rightarrow_4 \langle \mu(B, t_2), \nu(B, t_2)\rangle)$$

(where t_1 and t_2 are both maximal elements of T' for which the maximums of $\mu(A, t_1) - \nu(A, t_1)$ and of $\mu(B, t_2) - \nu(B, t_2)$ are achieved)

$$= \langle \max(\nu(A, t'), \mu(B, t')), \min(\mu(A, t'), \nu(B, t'))\rangle$$

$$\rightarrow \langle \max(\nu(A, t_1), \mu(B, t_2)), \min(\mu(A, t_1), \nu(B, t_2))\rangle$$

(where t' is the maximal element of T' for which the minimum of $\max(\nu(A, t'), \mu(B, t')) - \min(\mu(A, t'), \nu(B, t'))$ is reached)

$$= \langle \max(\nu(A, t_1), \mu(B, t_2)), \min(\mu(A, t'), \nu(B, t')),$$

$$\min(\mu(A, t_1), \nu(B, t_2), \max(\nu(A, t'), \mu(B, t'))\rangle.$$

Then, we consider the expression:

$$X = \max(\nu(A, t_1), \mu(B, t_2)), \min(\mu(A, t'), \nu(B, t'))$$

$$- \min(\mu(A, t_1), \nu(B, t_2), \max(\nu(A, t'), \mu(B, t')).$$

1. If for $\overline{t_2} \in T'$: $\mu(B, \overline{t_2}) \geq \nu(B, \overline{t_2})$, then:

$$X \geq \mu(B, \overline{t_2}) - \nu(B, \overline{t_2}) \geq 0.$$

2. If for $\overline{t_2} \in T'$: $\mu(B, \overline{t_2}) < \nu(B, \overline{t_2})$, then:

  2.1. If for $\overline{t_1} \in T'$: $\nu(A, \overline{t_1}) \geq \mu(A, \overline{t_1})$, then:

$$X \geq \nu(A, \overline{t_1}) - \mu(A, \overline{t_1}) \geq 0;$$

  2.2. If for $\overline{t_1} \in T'$: $\nu(A, \overline{t_1}) < \mu(A, \overline{t_1})$, then:

    2.2.1. If for $\overline{t'} \in T'$: $\min(\mu(A, t'), \nu(B, t')) \geq \max(\nu(A, t'), \mu(B, t'))$, then

$$X \geq \min(\mu(A, t'), \nu(B, t')) - \max(\nu(A, t'), \mu(B, t')) \geq 0,$$

    2.2.2 Otherwise, for $t' \in T'$: $\min(\mu(A, t'), \nu(B, t')) < \max(\nu(A, t'), \mu(B, t'))$, then

$$X \geq \min(\mu(A, t_0), \nu(B, t_0)) - \min(\mu(A, t_0), \nu(B, t_0)) = 0.$$

Therefore, in all cases $X \geq 0$, i.e., (a) is valid.

Assertions (b)–(d) are proved analogically. □

A geometrical interpretation of the temporal IFL is given on Fig. 4.1.

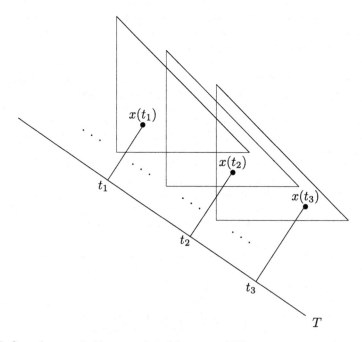

Fig. 4.1 Second geometrical interpretation of the temporal IFL

4.2 Multidimensional Intuitionistic Fuzzy Logics

Let $E, Z_1, Z_2, \ldots, Z_n$ be fixed finite linearly ordered sets.

By analogy with intuitionistic fuzzy multi-dimensional sets, introduced by E. Szmidt, J. Kacprzyk and the author in [4–7], here, following the paper [8] of the same authors and I. Georgiev, for a predicate P of the variables $x, z_1, z_2, \ldots, z_n$, ordered in the present form, we define an intuitionistic fuzzy evaluation function V for P in the form

$$V(P(x, z_1, z_2, \ldots, z_n)) = \langle \mu_P(x, z_1, z_2, \ldots, z_n), \nu_P(x, z_1, z_2, \ldots, z_n) \rangle,$$

where $x \in E$ is a (basic) variable, $z_1 \in Z_1, z_2 \in Z_2, \ldots, z_n \in Z_n$ are additional variables, $\mu_P(x, z_1, z_2, \ldots, z_n) \in [0, 1]$, $\nu_P(x, z_1, z_2, \ldots, z_n) \in [0, 1]$ and

$$\mu_P(x, z_1, z_2, \ldots, z_n) + \nu_P(x, z_1, z_2, \ldots, z_n) \leq 1.$$

Here, $\mu_P(x, z_1, z_2, \ldots, z_n)$ and $\nu_P(x, z_1, z_2, \ldots, z_n)$ are the degrees of validity and non-validity of $P(x, z_1, z_2, \ldots, z_n)$, respectively.

In the particular case, when $n = 1$, we obtain the case of temporal IFL (see [1]).

Having in mind the results from [4], we can define the following $(n + 1)$-dimensional intuitionistic fuzzy quantifiers:

(a) (partial) standard quantifier

$$V(\exists(x, z_1, z_2, \ldots, z_n)P(x, z_1, z_2, \ldots, z_n))$$

$$= \left\langle \max_{y \in E} \mu_P(y, z_1, z_2, \ldots, z_n), \min_{y \in E} \nu_P(y, z_1, z_2, \ldots, z_n) \right\rangle,$$

$$V(\forall(x, z_1, z_2, \ldots, z_n)P(x, z_1, z_2, \ldots, z_n))$$

$$= \left\langle \min_{y \in E} \mu_P(y, z_1, z_2, \ldots, z_n), \max_{y \in E} \nu_P(y, z_1, z_2, \ldots, z_n) \right\rangle$$

(b) (partial) i-quantifiers

$$V(\exists^i(x, z_1, z_2, \ldots, z_n)P(x, z_1, z_2, \ldots, z_n))$$

$$= \left\langle \max_{t_i \in Z_i} \mu_P(x, z_1, z_2, \ldots, z_{i-1}, t_i, z_{i+1}, \ldots, z_n), \right.$$

$$\left. \min_{t_i \in Z_i} \nu_P(x, z_1, z_2, \ldots, z_{i-1}, t_i, z_{i+1}, \ldots, z_n) \right\rangle,$$

$$V\left(\forall^i(x, z_1, z_2, \ldots, z_n) P(x, z_1, z_2, \ldots, z_n)\right)$$

$$= \left\langle \min_{t_i \in Z_i} \mu_P(x, z_1, z_2, \ldots, z_{i-1}, t_i, z_{i+1}, \ldots, z_n), \right.$$

$$\left. \max_{t_i \in Z_i} \nu_P(x, z_1, z_2, \ldots, z_{i-1}, t_i, z_{i+1}, \ldots, z_n) \right\rangle$$

(c) general additional quantifier

$$V\left(\exists^a(x, z_1, z_2, \ldots, z_n) P(x, z_1, z_2, \ldots, z_n)\right)$$

$$= \left\langle \max_{t_1 \in Z_1} \ldots \max_{t_n \in Z_n} \mu_P(x, t_1, t_2, \ldots, t_n), \min_{t_1 \in Z_1} \ldots \min_{t_n \in Z_n} \nu_P(x, t_1, t_2, \ldots, t_n) \right\rangle,$$

$$V\left(\forall^a(x, z_1, z_2, \ldots, z_n) P(x, z_1, z_2, \ldots, z_n)\right)$$

$$= \left\langle \min_{t_1 \in Z_1} \ldots \min_{t_n \in Z_n} \mu_P(x, t_1, t_2, \ldots, t_n), \max_{t_1 \in Z_1} \ldots \max_{t_n \in Z_n} \nu_P(x, t_1, t_2, \ldots, t_n) \right\rangle$$

(d) general quantifier

$$V\left(\exists^g(x, z_1, z_2, \ldots, z_n) P(x, z_1, z_2, \ldots, z_n)\right)$$

$$= \left\langle \max_{y \in E} \max_{t_1 \in Z_1} \ldots \max_{t_n \in Z_n} \mu_P(y, t_1, t_2, \ldots, t_n), \right.$$

$$\left. \min_{y \in E} \min_{t_1 \in Z_1} \ldots \min_{t_n \in Z_n} \nu_P(y, t_1, t_2, \ldots, t_n) \right\rangle,$$

$$V\left(\forall^g(x, z_1, z_2, \ldots, z_n) P(y, z_1, z_2, \ldots, z_n)\right)$$

$$= \left\langle \min_{y \in E} \min_{t_1 \in Z_1} \ldots \min_{t_n \in Z_n} \mu_P(y, t_1, t_2, \ldots, t_n), \right.$$

$$\left. \max_{y \in E} \max_{t_1 \in Z_1} \ldots \max_{t_n \in Z_n} \nu_P(y, t_1, t_2, \ldots, t_n) \right\rangle$$

Theorem 4.2.1 *For each of the five pairs of quantifiers, the following equalities hold:*

$$V(\neg\exists(x, z_1, z_2, \ldots, z_n)\neg P(x, z_1, z_2, \ldots, z_n))$$
$$= V(\forall(x, z_1, z_2, \ldots, z_n)P(x, z_1, z_2, \ldots, z_n)),$$
$$V(\neg\forall(x, z_1, z_2, \ldots, z_n)\neg P(x, z_1, z_2, \ldots, z_n))$$
$$= V(\exists(x, z_1, z_2, \ldots, z_n)P(x, z_1, z_2, \ldots, z_n)),$$
$$V(\neg\exists^i(x, z_1, z_2, \ldots, z_n)\neg P(x, z_1, z_2, \ldots, z_n))$$
$$= V(\forall^i(x, z_1, z_2, \ldots, z_n)P(x, z_1, z_2, \ldots, z_n)),$$
$$V(\neg\forall^i(x, z_1, z_2, \ldots, z_n)\neg P(x, z_1, z_2, \ldots, z_n))$$
$$= V(\exists^i(x, z_1, z_2, \ldots, z_n)P(x, z_1, z_2, \ldots, z_n)),$$
$$V(\neg\exists^a(x, z_1, z_2, \ldots, z_n)\neg P(x, z_1, z_2, \ldots, z_n))$$
$$= V(\forall^a(x, z_1, z_2, \ldots, z_n)P(x, z_1, z_2, \ldots, z_n)),$$
$$V(\neg\forall^a(x, z_1, z_2, \ldots, z_n)\neg P(x, z_1, z_2, \ldots, z_n))$$
$$= V(\exists^a(x, z_1, z_2, \ldots, z_n)P(x, z_1, z_2, \ldots, z_n)),$$
$$V(\neg\exists^g(x, z_1, z_2, \ldots, z_n)\neg P(x, z_1, z_2, \ldots, z_n))$$
$$= V(\forall^g(x, z_1, z_2, \ldots, z_n)P(x, z_1, z_2, \ldots, z_n)),$$
$$V(\neg\forall^g(x, z_1, z_2, \ldots, z_n)\neg P(x, z_1, z_2, \ldots, z_n))$$
$$= V(\exists^g(x, z_1, z_2, \ldots, z_n)P(x, z_1, z_2, \ldots, z_n)).$$

Proof Let us check the validity of the first equality.

$$V(\neg\exists(x, z_1, z_2, \ldots, z_n)\neg P(x, z_1, z_2, \ldots, z_n))$$

$$= \neg\exists(x, z_1, z_2, \ldots, z_n)\neg\langle\mu_P(x, z_1, z_2, \ldots, z_n), \nu_P(x, z_1, z_2, \ldots, z_n)\rangle$$

$$= \neg\exists(x, z_1, z_2, \ldots, z_n)\langle\nu_P(x, z_1, z_2, \ldots, z_n), \mu_P(x, z_1, z_2, \ldots, z_n)\rangle$$

$$= \neg\langle\max_{y\in E}\nu_P(y, z_1, z_2, \ldots, z_n), \min_{y\in E}\mu_P(y, z_1, z_2, \ldots, z_n)\rangle$$

$$= \langle\min_{y\in E}\mu_P(y, z_1, z_2, \ldots, z_n), \max_{y\in E}\nu_P(y, z_1, z_2, \ldots, z_n)\rangle$$

$$= V(\forall(x, z_1, z_2, \ldots, z_n)P(x, z_1, z_2, \ldots, z_n)).$$

The other equalities are proved in the same manner. □

An important problem arises.

Open Problem 21. Which other negations, different from the ones defined in Sect. 1.4, also satisfy these equalities?

For a finite linearly ordered set X, $i \in \{1, \ldots, 185\}$, $j \in \{1, 2, 3\}$ and $(n + 1)$-dimensional predicate P we define

$$\min_{x \in X}^{i,j} P(x, y_1, \ldots, y_n)$$

$$= P(x_1, y_1, \ldots, y_n) \wedge_{i,j} P(x_2, y_1, \ldots, y_n) \wedge_{i,j} \cdots \wedge_{i,j} P(x_m, y_1, \ldots, y_n),$$

$$\max_{x \in X}^{i,j} P(x, y_1, \ldots, y_n)$$

$$= P(x_1, y_1, \ldots, y_n) \vee_{i,j} P(x_2, y_1, \ldots, y_n) \vee_{i,j} \cdots \vee_{i,j} P(x_m, y_1, \ldots, y_n),$$

where $x_1 < x_2 < \cdots < x_m$ are the elements of X, listed in ascending order. Both operations produce a predicate of $y_1, \ldots, y_n$.

Let us define

$$\overline{Q}_{i,j} = \begin{cases} \exists_{i,j}, & \text{if } Q_{i,j} \text{ is } \forall_{i,j} \\ \\ \forall_{i,j}, & \text{if } Q_{i,j} \text{ is } \exists_{i,j} \end{cases}.$$

After these remarks, we continue with definitions of new quantifiers over the $(n + 1)$-dimensional predicate P. Let for $1 \leq i \leq 185$ and for $1 \leq j \leq 3$: $Q_{i,j} \in \{\forall_{i,j}, \exists_{i,j}\}$. Let

$$Q_{i,j}_{x \in X} = \begin{cases} \max_{x \in X}^{i,j}, & \text{if } Q_{i,j} = \exists_{i,j} \\ \min_{x \in X}^{i,j}, & \text{if } Q_{i,j} = \forall_{i,j} \end{cases},$$

Then, for $i, i_1, \ldots, i_n \in \{1, 2, \ldots, 185\}$ and $j, j_1, \ldots, j_n \in \{1, 2, 3\}$, we define:

(e) general Q-additional quantifiers

$$V((Q^1_{i_1,j_1}, Q^2_{i_2,j_2}, \ldots, Q^n_{i_n,j_n})(x, z_1, z_2, \ldots, z_n) P(x, z_1, z_2, \ldots, z_n))$$

$$= \left\langle Q^1_{i_1,j_1} \cdots Q^n_{i_n,j_n} \mu_P(x, t_1, t_2, \ldots, t_n), \overline{Q}^1_{i_1,j_1} \cdots \overline{Q}^n_{i_n,j_n} \nu_P(x, t_1, t_2, \ldots, t_n) \right\rangle,$$

(f) general Q-quantifier

$$V((Q_{i,j}, Q^1_{i_1,j_1}, Q^2_{i_2,j_2}, \ldots, Q^n_{i_n,j_n})(x, z_1, z_2, \ldots, z_n) P(x, z_1, z_2, \ldots, z_n))$$

$$= \left\langle Q_{i,j}_{y \in E} Q^1_{i_1,j_1}_{t_1 \in Z_1} \cdots Q^n_{i_n,j_n}_{t_n \in Z_n} \mu_P(y, t_1, t_2, \ldots, t_n), \right.$$

$$\left. \overline{Q}_{i,j}_{y \in E} \overline{Q}^1_{i_1,j_1}_{t_1 \in Z_1} \cdots \overline{Q}^n_{i_n,j_n}_{t_n \in Z_n} \nu_P(y, t_1, t_2, \ldots, t_n) \right\rangle,$$

Theorem 4.2.2 *For each of the three quantifiers, the following equalities hold:*

$$V(\neg_1(Q^1_{i_1,j_1}, Q^2_{i_2,j_2}, \ldots, Q^n_{i_n,j_n})(x, z_1, z_2, \ldots, z_n)\neg_1 P(x, z_1, z_2, \ldots, z_n))$$

$$= V((\overline{Q}^1_{i_1,j_1}, \overline{Q}^2_{i_2,j_2}, \ldots, \overline{Q}^n_{i_n,j_n})(x, z_1, z_2, \ldots, z_n) P(x, z_1, z_2, \ldots, z_n))$$

$$V(\neg_1(Q_{i,j}, Q^1_{i_1,j_1}, Q^2_{i_2,j_2}, \ldots, Q^n_{i_n,j_n})(x, z_1, z_2, \ldots, z_n)\neg_1 P(x, z_1, z_2, \ldots, z_n))$$

$$= V((\overline{Q}_{i,j}, \overline{Q}^1_{i_1,j_1}, \overline{Q}^2_{i_2,j_2}, \ldots, \overline{Q}^n_{i_n,j_n})(x, z_1, z_2, \ldots, z_n) P(x, z_1, z_2, \ldots, z_n)).$$

Open Problem 22. Which of the equalities from Theorems 4.2.1 and 4.2.2. are valid for the new quantifiers?

The so defined multidimensional intuitionistic fuzzy quantifiers can obtain different applications in the area of artificial intelligence. For example, we can use them in procedures for decision making and for intercriteria analysis, in rules of intuitionistic fuzzy expert systems, and others.

All these multidimensional intuitionistic fuzzy quantifiers are first-order.

The author believe that in a near future possibilities for defining second and higher-order multidimensional intuitionistic fuzzy quantifiers will arise and some properties for standard predicates, discussed in [9–20] will be studied for the multidimensional intuitionistic fuzzy quantifiers.

In future, we will study the possibility to change the condition "Let $E, Z_1, Z_2, \ldots, Z_n$ be fixed finite linearly ordered sets" with which Sect. 4.2 started. When the properties of the new intuitionistic fuzzy conjunctions and disjunctions are studied, probably, we will be able to change this condition with the condition "Let $E, Z_1, Z_2, \ldots, Z_n$ be fixed finite partially ordered sets". So, the new constructions will give additional possibilities for application in some areas of the artificial intelligence.

References

1. Atanassov K. Remark on a temporal intuitionistic fuzzy logic. Second Scientific Session of the "Mathematical Foundation Artificial Intelligence" Seminar, Sofia, March 30. Preprint IM-MFAIS-1-90, Sofia, 1990. p. 1–5, Reprinted: Int J Bioautomation. 2016;20(S1):S63-S68.
2. Karavaev E. Foundations of temporal logic. Leningrad: Leningrad University Publishing House; 1983 (in Russian).
3. Atanassov K. On intuitionistic fuzzy sets theory. Berlin: Springer; 2012.
4. Atanassov K, Szmidt E, Kacprzyk J. On intuitionistic fuzzy multi-dimensional sets. Issues Intuitionistic Fuzzy Sets Gen Nets. 2008;7:1–6.

5. Atanassov K, Szmidt E, Kacprzyk J, Rangasamy P. On intuitionistic fuzzy multi-dimensional sets. Part 2. In: Advances in fuzzy sets, intuitionistic fuzzy sets, generalized nets and related topics. Vol. I: Foundations. Warszawa: Academic. 2008. P. 43–51.
6. Atanassov K, Szmidt E, Kacprzyk J. On intuitionistic fuzzy multi-dimensional sets. Part 3. In: Developments in fuzzy sets, intuitionistic fuzzy sets, generalized nets and related topics, Vol. I: Foundations. Warsaw: SRI Polish Academy of Sciences. 2010. P. 19–26.
7. Atanassov K, Szmidt E, Kacprzyk J. On intuitionistic fuzzy multi-dimensional sets. Part 4. Notes Intuitionistic Fuzzy Sets. 2011;17(2):1–7.
8. Atanassov K, Georgiev I, Szmidt E, Kacprzyk J. Multidimensional intuitionistic fuzzy quantifiers. In: Proceedings of the 8th IEEE Conference "Intelligent Systems", Sofia, 4–6 September 2016 pp. 530–534.
9. Barwise J, editor. Handbook of mathematical logic., Studies in Logic and the Foundations of MathematicsAmsterdam: North Holland; 1989.
10. Crossley JN, Ash CJ, Brickhill CJ, Stillwell JC, Williams NH. What is mathematical logic?. London: Oxford University Press; 1972.
11. van Dalen D. Logic and structure. Berlin: Springer; 2013.
12. Ebbinghaus H-D, Flum J, Thomas W. Mathematical logic. 2nd ed. New York: Springer; 1994.
13. Lindstrm P. First-order predicate logic with generalized quantifiers. Theoria. 1966;32:186–195.
14. Mendelson E. Introduction to mathematical logic. Princeton: D. Van Nostrand; 1964.
15. Mostowski A. On a generalization of quantifiers. Fund Math. 1957;44:12–36.
16. Mostowski M. Computational semantics for monadic quantifiers. J Appl Non-Class. Logics. 1998;8:107–121.
17. Shoenfield JR. Mathematical logic. 2nd ed. Natick: A.K. Peters; 2001.
18. http://www.cl.cam.ac.uk/~aac10/teaching-notes/gq.pdf
19. http://plato.stanford.edu/entries/generalized-quantifiers/
20. http://www.unige.ch/lettres/linguistique/files/5014/1831/0318/MMil_ch_14_GQs.pdf

Chapter 5
Conclusion

In the book, the author has collected his basic results related to intuitionistic fuzzy logics. He hopes that the book will motivate the fuzzy and intuitionistic fuzzy community for future research and development of this area of the Zadeh's fuzzy world.

The author's plans for the near future are related to the extensions of the intuitionistic fuzzy sets and logics, such as interval-valued intuitionistic fuzziness, intuitionistic L−fuzziness, intuitionistic fuzziness of type 2 and n, etc.

During last years, some new types of sets, that are completely identical with the intuitionistic fuzzy sets were introduced, under other names. The author cannot see anything positive in that: practically, these attempts do not develop the existing research over intuitionistic fuzziness; they only increase the terminological chaos. As it is seen in the book, now, there is a long list of open problems that are waiting for their solutions. Therefore, instead of reinventing the wheel and introducing older objects under new names, it will be better if we all work on solving these open problems, and formulate new ones.

This being said, the author is optimist about the future of the intuitionistic fuzziness.

© Springer International Publishing AG 2017
K.T. Atanassov, *Intuitionistic Fuzzy Logics*, Studies in Fuzziness and Soft Computing 351, DOI 10.1007/978-3-319-48953-7_5

Index

A
Angelova, N., 32
Atanassova, L., 10, 27
Atanassova, V., 5
Axiom of resolution, 41
Axioms
    of conditional logic, 42
    of intuitionistic logic, 30
    of Klir and Yuan, 32
    of Kolmogorov, 31
    of Łukasiewicz–Tarski, 31
    of the $\mathcal{K}$-theory, 66

B
Brower, L., 2

C
Çuvalcioğlu, G., 111

D
Dencheva, K., 108
Dworniczak, P., 5, 10

G
Geometrical interpretation, 3, 5, 65, 72, 79, 107

I
Intuitionistic fuzzy negation, 20
Intuitionistic fuzzy pair, 1
Intuitionistic fuzzy sure, 8
Intuitionistic fuzzy tautology, 7

K
Kacprzyk, J., 10, 125
Klir, G., 32, 59
Kolev, B., 9
Kolmogorov, A., 31

L
Law
    De Morgan, 55
    for excluded middle, 52
    for excluded middle, modified, 53
    Hauber, 39
    of contraposition, 32
Logical falsity, 2
Logical truth, 1
Łukasiewicz, J., 31

M
Meredith, C. A., 28
Multi unitary group, 59

O
Open problem, 26, 27, 41, 52, 59, 60, 71, 74, 76, 82, 84, 96, 99, 100, 113, 120, 131, 133
Operation
    conjunction, 3, 58
    disjunction, 3, 58
    implication, 3, 7, 10, 19, 32
    negation, 2, 19
Operator
    extended modal operator
        D_α, 93
        $F_{\alpha,\beta}$, 93

© Springer International Publishing AG 2017
K.T. Atanassov, *Intuitionistic Fuzzy Logics*, Studies in Fuzziness
and Soft Computing 351, DOI 10.1007/978-3-319-48953-7

$G_{\alpha,\beta}$, 93
$H_{\alpha,\beta}$, 93
$J_{\alpha,\beta}$, 93
$X_{\alpha,\beta,\gamma,\delta,\varepsilon,\eta}$, 93
level operator
   $P_{\alpha,\beta}$, 120
   $Q_{\alpha,\beta}$, 120
modal operator
   necessity, $\square$, 79
   possibility, $\diamondsuit$, 79
   $\bigcirc$, 90
second-type of modal operator
   $\boxplus$, 106
   $\boxtimes$, 106
   $\boxplus_\alpha$, 107
   $\boxtimes_\alpha$, 107
   $\boxplus_{\alpha,\beta}$, 108
   $\boxtimes_{\alpha,\beta}$, 108
   $\boxplus_{\alpha,\beta,\gamma}$, 109
   $\boxtimes_{\alpha,\beta,\gamma}$, 109
   $\boxdot_{\alpha,\beta,\gamma,\delta}$, 110
   $\boxcircle_{\alpha,\beta,\gamma,\delta,\varepsilon,\zeta}$, 112
   $E_{\alpha,\beta}$, 111
temporal operator
   F, 126
   G, 126
   H, 126
   P, 126
weight-center operator, W, 76

Q
Quantifier
   extended IF quantifier
      for all ($\forall_\mu$), 71
      for all ($\forall_\nu$), 71
      for all ($\forall_\nu^*$), 72
      there exists ($\exists_\mu$), 71
      there exists ($\exists_\nu$), 71
      there exists ($\exists_\nu^*$), 72
   multidimensional IF
      general additional quantifier, 130
      general Q-additional quantifier, 132
      general Q-quantifier, 133
      general quantifier, 130
      (partial) i-quantifier, 130
      (partial) standard quantifier, 129
   standard IF quantifier
      for all, $\forall$, 65
   standard IF quantifier
      there exists, $\exists$, 65

R
Rose, G.F., 19
Rule of conditional logic, 42

S
Szmidt, E., 10, 125

T
Tarski, A., 31
Tautology, 7
Trifonov, T., 9

Y
Yuan, B., 32, 59

Z
Zadeh, L., 10

Printed in the United States
By Bookmasters